The MACAT Library

世界思想宝库钥匙丛书

解析艾尔弗雷德·W.克罗斯比

《哥伦布大交换：1492年以后的生物影响和文化冲击》

AN ANALYSIS OF

ALFRED W. CROSBY'S

THE COLUMBIAN EXCHANGE

Biological and Cultural Consequences of 1492

Joshua Specht　Etienne Stockland ◎ 著

宫昀 ◎ 译

上海外语教育出版社
SHANGHAI FOREIGN LANGUAGE EDUCATION PRESS

目　录

CONTENTS

引　言

要 点

- 艾尔弗雷德·W.克罗斯比是美国环境史学家，德克萨斯大学奥斯汀分校荣休教授。

- 在《哥伦布大交换》中，克罗斯比认为，欧洲与美洲之间植物、动物、疾病的迁移，是15世纪意大利探险家克里斯托弗·哥伦布 * 的航海之旅带来的最重要成果，他也被认为是欧洲殖民史上的关键人物。

- 《哥伦布大交换》指出，在欧洲对美洲的殖民进程中，生态因素起着重要作用。

艾尔弗雷德·W.克罗斯比其人

艾尔弗雷德·W.克罗斯比，《哥伦布大交换：1492年以后的生物影响和文化冲击》（1972）的作者，1931年出生于马萨诸塞州波士顿市，1952年毕业于哈佛大学，1952—1955年在巴拿马服兵役。退役后，他在哈佛大学教育学院获教学艺术硕士学位，1961年又在波士顿大学获历史学博士学位，期间师从著名美国殖民史学家罗伯特·E.穆迪 *。克罗斯比的博士论文《美国、俄国、大麻和拿破仑》，作为他的首部著作于1965年发表。该书追溯了1783—1812年间，美国与波罗的海地区贸易的经济史。

然而，克罗斯比迅速成为了当时主流传统历史研究的批评者，他的作品也受益于该时期政治和智识的发展。美国生物学家雷切尔·卡森 * 的力作《寂静的春天》（1962），主要探讨了农药使用的潜在生态影响。自其出版以来，基于环境保护的行动主义事业蒸蒸日上。与之类似地，克罗斯比在2003版《哥伦布大交换》序言中称，他这本书缘起于 20 世纪 60 年代的社会动荡，主要受美国民权运动 *、黑人权力运动 *（主要是美国黑人为争取民权和政治平等而斗争）和越南战争 *（美国支持的南越与信仰共产主义的北越间的冲突，造成了数万名美国人丧生）的启发。

前辈历史学家只关心欢庆的政治史书写。在序言中，克罗斯比提到，"对这些人来说……好人总是获胜。"[1] 但 20 世纪 60 年代的动乱，使他对此前的历史叙述心生怀疑，他开始思考除政治以外的接近历史的路径，这才有了后来他在科学和医学人类学领域的研究。

在职业生涯中，克罗斯比曾短暂任教于俄亥俄州立大学和耶鲁大学。他大部分时间都在德克萨斯大学奥斯汀分校工作，但《哥伦布大交换》的研究和撰写完成于华盛顿州立大学。2015 年，克罗斯比作为历史学、地理学和美国研究教授，从德克萨斯大学奥斯汀分校荣誉退休。

《哥伦布大交换》的主要内容

早期历史学家认为，1492 年，在克里斯托弗·哥伦布实现从欧洲到美洲的航行（这次探险被认为对欧洲殖民主义有重大影响）之后，历史主要由社会、政治因素驱动。然而在《哥伦布大交换》中，克罗斯比坚称，生物因素对历史变化的影响尤为重要。他勾勒出一个"哥伦布大交换"，认为从前分处异地的植物、动物、微生物和人相互接触，导致全球生物走向同质化 *。换句话说，随着一些植物和动物散布全世界，它们的竞争者就会消失或被边缘化。在解释欧洲殖民在美洲的成功之路时，克罗斯比谈到生物迁移，例如向欧洲输入美洲的高热量食物，如马铃薯、树薯（又名木薯）

和玉米。

在考察 1492 年航海后欧洲的人口 * 大爆炸（即人口数量激增）时，克罗斯比指出，人口大爆炸的原因在于美洲粮食作物在欧洲的传播。[2] 他论证道，观察每公顷作物产出的百万计卡路里可发现，美洲作物的产出远高于旧世界 *（非洲、欧洲和亚洲）的作物。玉蜀黍（玉米）、马铃薯和红薯（番薯）每公顷产出超过 700 万卡路里，旧世界作物只有大米能达到 730 万卡路里，而南美本地作物木薯的产出是 990 万。这些食物成为欧洲和亚洲饮食的重要组成部分，其中木薯和玉米则推动了非洲的人口增长（如今已成为他们的主食）。另一方面，克罗斯比指出，输入到美洲的牲畜，使西半球能够供养比以前更多的人口。新世界 * 作物的传播使欧洲人口迅速增长，这些人口随后又通过移民来到新世界（即美洲和加勒比群岛）。

早期的著作认为，政治和技术因素，如更先进的武器，帮助西班牙人征服了新世界。克罗斯比则认为，传染性病毒天花 *，在从前未接触过该病毒的人群中疯狂肆虐，才是新世界人民向征服者 *（西班牙征服者）缴械投降的真正元凶。他追踪微生物病害对社会结构和政治组织的影响，并总结道，传染病造成的间接破坏（造成经济和社会崩溃），和直接破坏（造成人口死亡）一样具有毁灭性。例如，他发现土著印加人在权力斗争中抵抗力减弱，他们的统治者怀纳·卡帕克 * 突然死亡，几乎可以肯定元凶是疾病。克罗斯比把这一点与他的更大主题相联系，强调在生态背景下了解人类历史的重要性。在《哥伦布大交换》的每一章节中，他考察哥伦布航海后历史的一个不同方面，并同时指出，与社会和政治因素比起来，生物力量能更好地解释历史现象。

《哥伦布大交换》的学术价值

在搜集论据和展开论证两方面，《哥伦布大交换》都堪称过去50年间最重要的历史著作之一。虽然历史书写主要围绕人类事件和活动而展开，克罗斯比的论证却极具说服力地指出，1492年克里斯托弗·哥伦布的航海带来的最重大结果都是非人为因素造成的。这个观点如今已广为接受，但在该书首次出版的1972年，它却是具有革命性意义的。

此外，对如何借助科学进行历史研究，《哥伦布大交换》起到了绝佳的示范作用。在接触当时的科学文献时，克罗斯比表现出极高的技巧和创造性，这表明这些领域可以给史学家教益，史学家也能够在流行病学*（流行疾病的动态研究）、生物学和其他科学领域贡献重要观点。因此，克罗斯比的著作值得持续关注，它不仅是重要的史学文本，是环境史领域的奠基著作，也是跨学科研究（即从若干其他领域借鉴方法和目标的研究）的绝佳范例。

《哥伦布大交换》的中心观点被证实具有极大影响力，它认为，是流行疾病帮助欧洲人征服了新世界。早先的学者都将这场征服归因于欧洲的技术，或美洲人民"原始的"宗教和政治制度。通过阐明疾病的影响，克罗斯比推翻了这些陈旧的叙述，证明了生物因素的重要性。克罗斯比的"哥伦布大交换"，已成为构建特定时代历史的重要框架。该文本的影响力经久不衰，也因此成为最富意义的历史著作之一。

关于克里斯托弗·哥伦布1492年从欧洲到美洲的航行，《哥伦布大交换》彻底重构了此前学者的认识，有力证明了环境史学在各历史学科中的重要地位。凭借这些成就，该著作全面打开了历史研究的新领域。

1　艾尔弗雷德·W. 克罗斯比：《哥伦布大交换：1492 年以后的生物影响和文化冲击》，康涅狄格州韦斯特波特：普雷格出版社，2003 年，第 xx 页。

2　克罗斯比：《哥伦布大交换》，第 165—207 页。

第一部分：学术渊源

1 作者生平与历史背景

要点 🔑

- 美国环境史学家艾尔弗雷德·W.克罗斯比的著作，论证了植物、动物、疾病等非人为因素影响历史的过程。
- 克罗斯比成为"进步叙事"的批评者，这种叙事是 20 世纪中叶主流的历史书写方式，它认为历史研究就是记录人类的进步。
- 克罗斯比撰写《哥伦布大交换》时，正值 20 世纪 60、70 年代环保主义运动高涨。

为何要读这部著作？

艾尔弗雷德·W.克罗斯比 1972 年出版《哥伦布大交换：1492 年以后的生物影响和文化冲击》，该书认为，"哥伦布航海带来的最重要改变是生物性的。"[1] 克罗斯比从意大利探险家克里斯托弗·哥伦布 1492 年的航海入手，考察东、西半球交流的潜在影响因素。早先的研究强调文化和技术因素对东、西半球关系的影响，而克罗斯比则认为，旧世界和新世界间植物、动物、微生物的交换等非人为因素影响力更大。

克罗斯比的观点受到一些早期作品影响，如美国历史学家沃尔特·普雷斯科特·韦伯 * 的《大平原》（1931），它探索了特定的环境和生态系统。[2] 克罗斯比在这些作品中寻找思路，形成整体性 * （综合的和普遍的）观点，考察非人为的生物性因素如何决定性地塑造和改变人类历史，哥伦布到达新大陆之后的情况验证了这一点。《哥伦布大交换》是环境史学科几部奠基著作之一，或许它为环境史学家提供了最有效的范例，告诉他们如何从地理学、人类学等学科中汲取灵感，重构关键历史时刻的流行观点。

> "20世纪70年代是令环保人士欢欣鼓舞的时代，这10年间美国确立了第一个地球日，筹建了环保总局。在接下来的几年陆续出台了《清洁水法案》《濒危物种法案》和《环境农药控制法案》。同时，环境历史学逐渐成为专门和独立的研究学科。"
>
> ——艾尔弗雷德·W. 克罗斯比，"环境史的过去与现在"

作者生平

克罗斯比本科就读于哈佛大学历史学专业。1961年在波士顿大学获历史学博士学位，师从美国历史学家罗伯特·E. 穆迪。尽管接受的是正统教育，但克罗斯比迅速成为主流历史研究的批评者。在2003版《哥伦布大交换》中，克罗斯比提到他的老师们从不怀疑社会的基本的善，即他们在二战中为之奋斗的目标。他们相信世界正逐步变好，"好人一定会赢"。[3] 但对克罗斯比来说，20世纪60年代的动乱使他对传统叙事产生怀疑，他开始思考除政治以外的探索历史的路径，这才有了后来他借助科学和医学人类学做研究的经历。克罗斯比在《美国人类学家》和《拉丁美洲历史评论》上发表两篇论文后，开始着手《哥伦布大交换》的撰写，探讨哥伦布航海带来的后果。克罗斯比曾在多所大学任教，包括俄亥俄州立大学、耶鲁大学、赫尔辛基大学，以及他筹备和撰写《哥伦布大交换》的华盛顿州立大学等。2015年，克罗斯比以历史学、地理学和美国研究教授的身份，从德克萨斯大学奥斯汀分校荣誉退休，在那里他度过了职业生涯的大部分时间。

创作背景

　　20 世纪 60 年代的政治环境和智识氛围对克罗斯比的生活和工作产生了深远影响。1962 年，美国海洋生物学家雷切尔·卡森出版《寂静的春天》，探讨农药使用造成的生态后果，随后环保主义作为一项社会运动迅速兴起。当时民权运动（在此期间，美国黑人要求依法享有平等权利）、越南战争以及大规模的动乱纷纷出现，动摇了克罗斯比对西方文化优越性的信心，也使他质疑历史就是进步的问题这一观点。[4] 因此，克罗斯比开始转向历史以外的其他领域，谋求解决历史问题的方法。他的巨大影响力遍布生物学和其他一系列科学学科，如人类学（人类文化研究）、地理学、考古学、农学（农业科学，包括土壤管理）、生态学等多个相关领域。他认为，20 世纪 60 年代把一些人推向了教条的政治立场。他简单解释道，"20 世纪 60 年代使一些人成了理论家，却把我引入了生物学领域。"[5]

　　克罗斯比还研究流行病学，探讨疾病和食物在人类历史上的重要作用。一旦沉浸在这些话题中，他就放弃了思考构成历史学科应该考虑的问题，转而思考一些全然不同的、更为宏大的问题。例如，他会考虑，"是什么让人类活到足以能够繁殖后代？又是什么让他们死去？"[6] 当努力回答此类问题时，他认识到人类历史上非人为因素的决定性影响，并开始以全新的视角审视历史事件的探索。

1 艾尔弗雷德·W.克罗斯比:《哥伦布大交换:1492 年以后的生物影响和文化冲击》,康涅狄格州韦斯特波特:普雷格出版社,2003 年。

2 沃尔特·普雷斯科特·韦伯:《大平原》,林肯:内布拉斯加大学出版社,1981 年。详见艾尔弗雷德·W.克罗斯比:"环境史的过去和现在",《美国历史评论》100 卷,1995 年第 4 期,第 1177—1189 页。

3 克罗斯比,《哥伦布大交换》,第 xx 页。

4 克罗斯比:《哥伦布大交换》,第 xx—xxi 页。

5 克罗斯比:《哥伦布大交换》,第 xxi 页。

6 克罗斯比:《哥伦布大交换》,第 xxi 页。

2 学术背景

要点 ⚷

- 艾尔弗雷德·W.克罗斯比把历史学家的注意力从政治转移到了人类与环境的关系史上。

- 克罗斯比运用科学家使用的工具，研究了在面临环境压力时，人类社会如何生存和灭亡。

- 克罗斯比帮助史学家和科学家进行了交流和对话。

著作语境

1972年，艾尔弗雷德·W.克罗斯比出版了《哥伦布大交换：1492年以后的生物影响和文化冲击》，他的历史研究方法具有革命性。在此之前，传统史学著作重点关注具体国家的政治历史，人类始终居于历史研究的中心位置，克罗斯比期望改变这种定位。在研究中他认为，一系列更为广泛的生态关系决定和改变了人类历史。1492年，意大利探险家克里斯托弗·哥伦布从欧洲到美洲的航行带来了各种剧变。克罗斯比探究其中的非人为因素，并依此建立了自己的理论。

之所以采用这种方法，是因为克罗斯比希望解决困扰早期历史学家的恼人问题。在西班牙征服＊新世界的历史中，前人仅强调西班牙在政治、文化和技术上的优势，却未能据此解释中南美洲阿兹特克人＊和印加人人口骤减的原因。被克罗斯比置于西班牙征服史事件中心的，是流行性疾病，尤其是急性病毒性疾病天花。他运用有说服力的原创性叙事方式，合理解释了美洲土著文化中，社会和政治大规模迅速崩塌的原因。

借鉴了几个不同研究领域的目标和方法,《哥伦布大交换》的跨学科性奠定了克罗斯比成功的基础,也造就了它革命性的研究路径。然而这本书太特殊,最初克罗斯比竟找不到出版社愿意出版它。[1] 20 世纪 70 年代早期,历史著作更多还是从社会科学、人文学科中汲取养料。如今我们普遍认为,历史学和硬科学之间可以有重要对话。但在克罗斯比最初出版《哥伦布大交换》时,情形并非如此。因此该书改变了人们的长期观念,具有非凡意义。

> "我避开对历史的意识形态解读,转而探寻一些基本命题,比如生存和死亡……是什么让人类活到足以能够繁殖后代?又是什么让他们死去?或许是食物和疾病?"
>
> ——艾尔弗雷德·W.克罗斯比,2003 版《哥伦布大交换》序言

学科概览

在环境史这一新兴领域中,克罗斯比借鉴了前人的方法,但在抱负和视野上,他的著作与前人的并不相同。1893 年,美国历史学家弗雷德里克·杰克逊·特纳*发表了创新性的论文——"美国历史中边疆的意义",探讨美国的边疆环境如何帮助促进个人主义和民主。40 年后,历史学家沃尔特·普雷斯科特·韦伯发表《大平原》,研究特定环境和生态系统的变化。[2] 尽管韦伯的著作视野广阔,但《哥伦布大交换》前的大多数研究仍有其局限性。杰克逊和韦伯的著作主要涉及一般性环境主题,而非应用生物、生态等科学方法阐释人与非人类世界的关系。克罗斯比从这些著作中汲取养分,将他们的方法扩充成整体性视野,研究

植物、动物、微生物如何以多种方式，对人类历史产生决定性影响。

为实现这一目标，克罗斯比在地理学、生物学等多个领域寻求灵感，他或许是首个把这些学科带进历史研究的人，也是他把人类学和流行病学带进了历史书写的沃土。如此一来，《哥伦布大交换》提出了被其他历史学家忽视的问题，给出了他们从未思考过的答案。在 2003 版序言中，克罗斯比打了个比方，他说，提这样的问题就像"取出相机中的标准胶片，换上红外或紫外胶片"，所以才能够"看到从前看不见的东西"。[3]

学术渊源

克罗斯比的写作，受历史专业本身影响不大，主要是受到科学学科的影响，他才发现了全新的景象。在一篇题为"环境史的过去与现在"（1995）的文章中，克罗斯比引用了若干领域的著作，以突显它们对他的思想和整个环境史学科的重要作用。他提到，考古学家运用新技术，研究古代气候和生态系统；弗雷德里克·E. 克莱门茨*和查尔斯·埃尔顿*等生态学家拓宽了他的视野；保罗·维达尔·白兰士*等地理学家发展出一些理论，这些理论"为对环境展现浓厚兴趣的历史学家们带去了灵感"。例如，白兰士可能论*的论点是，环境并不完全决定人类文化和社会的本质。它取代了简单的环境决定论（即认为环境决定一切），使历史学家能在自然环境的有限范围内研究人类社会的发展。流行病学的进步也影响了克罗斯比的著作。对他来说，环境史学家乐意应用硬科学，会使整个领域变得不同，这也正是他极力推崇的。

克罗斯比总结变革时谈到，"环境史学家发现，物理和生命科学可以提供无数有用的信息和理论，它们对于历史调查甚至作用关键，科学家们也因此常常能够清晰表达自己的观点。"[4]

1 艾尔弗雷德·W.克罗斯比："重估 1492"，《美国季刊》41 卷，1989 年第 4 期，第 661 页。

2 沃尔特·普雷斯科特·韦伯：《大平原》，林肯：内布拉斯加大学出版社，1981 年。详见艾尔弗雷德·W.克罗斯比："环境史的过去和现在"，《美国历史评论》100 卷，1995 年第 4 期，第 1177—1189 页。

3 艾尔弗雷德·W.克罗斯比：《哥伦布大交换：1492 年以后的生物影响和文化冲击》，康涅狄格州韦斯特波特：普雷格出版社，2003 年，第 xxi 页。

4 克罗斯比："环境史的过去和现在"，第 1189 页。

3 主导命题

要点 🗝

- 在《哥伦布大交换》中，艾尔弗雷德·W.克罗斯比试图对欧洲在新世界殖民的"成功"做出解释。
- 历史学家过去常强调卓越的技术，以此来解释欧洲人如何征服新世界的美洲原住民。
- 克罗斯比强调了流行病的重要性，它对欧洲殖民统治起着媒介作用。

核心问题

在《哥伦布大交换：1492年以后的生物影响和文化冲击》中，艾尔弗雷德·W.克罗斯比的主要目标是对历史研究中的人为因素去中心化。事实上，正是由于对非人为因素的思考，他才能够解释，1492年，克里斯托弗·哥伦布的航行贯通新世界和旧世界之后，欧洲人口快速增长的原因。前人的研究将城市化、技术进步和经济创新视为关键原因。克罗斯比则指出，起作用的更重要因素是高热量的新世界粮食作物向欧洲的传播。

克罗斯比还认为，1492年以后，全球生物种类呈现同质性趋势（即物种多样性下降），这比新的跨大西洋贸易路线的开放更加重要。哥伦布航海之后，从前分隔开的人、植物和动物相互接触，在接触过程中，某些种群被大量摧毁。多样化的新世界文化，部分地被帝国主义*殖民地统治所取代，植物和动物的生物多样性让位于棉花、甘蔗、烟草等作物的大规模单一栽培（即单一作物的种植）。克罗斯比将这一趋势描述为，"自大陆冰川消融以来，本星球生命历史上最重要的篇章之一。"[1]

> "生物种类呈同质性趋势，这是自大陆冰川消融以来，本星球生命历史上最重要的篇章之一。"
>
> ——艾尔弗雷德·W. 克罗斯比，《哥伦布大交换》

参与者

早先的历史学家认为，在克里斯托弗·哥伦布 1492 年航海后的时代，社会和政治因素推动了历史。一个著名的例子是，比利时历史学家查尔斯·韦林登 * 将新世界欧洲帝国主义的根源追溯到十字军东征时期（中世纪欧洲国家组织军事远征力量向中东开进，目的是从穆斯林手中"重获"基督教圣地）的地中海历史，他认为当时"初次应用的组织结构和剥削手段，到 16 世纪被强加在了美洲人头上"。[2] 法国历史学家皮埃尔·乔努 * 同样认为，在地中海产糖岛屿上发展起来的政治和经济剥削手段，为在美洲的殖民统治铺平了道路。

与之截然不同的是，克罗斯比在《哥伦布大交换》中坚持认为，变化的最重要因素是生物性的。他勾勒出"哥伦布大交换"的生物关系网，指出从前分隔开的植物、动物、微生物开始和不同人群产生接触，造成全球生物同质性的整体趋势。此前的著作都援引了政治和技术因素（包括更好的武器）作为西班牙人征服新世界的主要推手，克罗斯比却提出全然不同的观点。他认为，天花在以前未接触过该病毒的人群中肆虐传播，是新世界人民不得不向征服者低头的更关键因素。

当代论战

克罗斯比的《哥伦布大交换》也曾遭到抨击，尤其是那些在历史叙述中强调社会文化因素重要性的学者，他们都对该书中的观点持有异议。在 1987 年的著作《哥伦布航海、哥伦布大交换和历史学家们》中，克罗斯比评论称，他研究欧洲殖民的路径"隐秘又令人不安"，尤其困扰那些持"欧美种族中心主义"[3]立场的人（即在论证中自动奉欧美视角为中心的人）。

克罗斯比的生态学方法遭遇的最强有力反对者是美国经济史学家大卫·兰德斯 *，他认为欧洲殖民的"成功"还是要归于政治、社会和文化因素。[4]兰德斯等评论者都强调技术变革的重要性，并观察到哥伦布航海本身首先依赖于社会、政治和技术力量，跨大西洋航行才成为可能，这类反响构成了对克罗斯比著作的关键性批评。尽管主流学术思想已转向克罗斯比的观点，但兰德斯和与他志同道合的历史学家的批评，仍是有力的少数派反对意见。随后几十年，包括理查德·怀特 * 和威廉·克罗农 * 在内的环境史学家同样认为，克罗斯比夸大了历史进程中生物因素的作用。这些史学家虽然也认可非人为因素的重要性，但他们认为这些因素与人类文化不能完全分开。

---

1 艾尔弗雷德·W. 克罗斯比：《哥伦布大交换：1492 年以后的生物影响和文化冲击》，康涅狄格州韦斯特波特：普雷格出版社，2003 年，第 3 页。

2　克罗斯比:《哥伦布大交换》, 第4—5页。

3　艾尔弗雷德·W. 克罗斯比:《哥伦布航海, 哥伦布大交换和历史学家们》, 华盛顿特区: 美国历史协会, 1987年, 第1—2页。

4　大卫·S. 兰德斯:《国富国穷: 为什么有的国家富有的国家穷》, 纽约: 诺顿出版社, 1998年。

4 作者贡献

要点 ⚿—

- 在《哥伦布大交换》中，艾尔弗雷德·W.克罗斯比试图阐明，粮食作物、动物、疾病等非人为因素的历史重要性。

- 克罗斯比对生态因素的关注让人们以新的方式看待欧洲殖民 * 在新世界的成功。

- 通过借鉴科学界的研究成果揭示历史问题，克罗斯比发展出一种全新的历史研究方法。

作者目标

在《哥伦布大交换：1492 年以后的生物影响和文化冲击》一书中，艾尔弗雷德·W.克罗斯比试图将人文学科、社会科学和生物科学的研究广泛结合，以期能够证实，"哥伦布航海后的最重要变化是生物因素造成的"。[1] 此前的大部分著作，都强调人为因素带来的重大历史改变，如西班牙人对阿兹特克人（今墨西哥所在地的原住民）的征服，或 1492 年以后欧洲经济的迅速增长。然而，克罗斯比展现了这些进程中，作为关键因素的生物迁移，包括土豆、木薯、玉米等高热量食物向欧洲的输送，以及天花等疾病向美洲的传播带来的流行病效应。

克罗斯比的更大意图在于，展示非人为的环境因素在理解人类历史时的关键作用。他认为，早先的历史学家过分强调政治和经济因素，他们未能认识到，只有在生物学语境下才能理解人类。为澄清这一点，他论述道，病原体 *（引起疾病的微生物）和生物群（特定地区的植物群和动物群）的意外迁移，同刻意的人类活动一样，共同促成了欧洲帝国在美洲的成功。在物质和观念两个方面，克罗斯比都期望证实，非人为力量才是人类历史的主要推手。

> "处女地流行病给人群带来危险，这些人之前从未接触过侵害他们的各种疾病，因此毫无免疫能力。这些流行病，如天花、麻疹、疟疾、黄热病及其他疾病，在美洲历史上非常重要。但在前哥伦布时代的新世界，它们是闻所未闻的。"
>
> ——艾尔弗雷德·W. 克罗斯比，"处女地流行病作为美洲土著人口减少的原因"

研究方法

《哥伦布大交换》的知识框架有赖于克罗斯比对人类学、地理学等相关领域著作的深入思考，以及对生物学、地质学和医学等硬科学领域成果的了解。克罗斯比援引的理论中影响力较大的，来自美国植物进化学者弗雷德里克·E. 克莱门茨和英国动物学家查尔斯·埃尔顿，他们研究的是入侵物种对自然生态系统的影响。克罗斯比的中心论点是，1492 年克里斯托弗·哥伦布的航海，对美洲的最大影响是生物性的。他尝试将历史学科以外的文本综合纳入他的历史研究著作中，并得出上述结论。

关于生物因素的中心作用，克罗斯比在得出几个较小结论的基础上，提出了自己的主要观点。事实上，在《哥伦布大交换》看似各无关联的章节中，生物主题比比皆是，贯通各个章节，每一章探讨东、西半球接触后的某一个环境后果。这一方法非常奏效，尽管之后的部分研究对克罗斯比的观点有所质疑，他仍实现了他的整体目标。他认为，哥伦布大交换将性传播疾病梅毒＊带到了旧世界，但这一点如今仍存争议。因为综合各种证据显示，这种疾病在欧洲或许本已存在。尽管如此，一个更大、更具说服力的事实是：如果综合观察人类学、地理学、生物学和其他学科领域的证据，会发现非人为因素才是哥伦布大交换带来的变化的决定性因素。

时代贡献

在《哥伦布大交换》的关键议题中，克罗斯比应用科学研究来回答具体的历史问题。在某种程度上，对于历史变革中非人为因素的强调，他立场积极，这是受到 20 世纪 70 年代初兴起的环保主义运动的启发。如今这一方法无疑属于环境史学，但在 1972 年克罗斯比撰写《哥伦布大交换》时，环境史作为一门独立的分支学科并不存在。

认定克罗斯比在《哥伦布大交换》中的观点完全原创并不正确。但是，尽管他在历史研究中借用了科学学科中的某些已有观点，他的结论仍然是高度创新的。他认为科学研究可以启发历史研究的观点非常新潮。因为在此之前，1492 年哥伦布从欧洲到美洲的航海史，主要受政治因素支配。此外，克罗斯比又将此反转，认为历史也能启发科学，但这一观点如今还是像 1972 年时一样界限不明。自《哥伦布大交换》以后，历史学家们都努力地广泛接触社会科学和人文学科，但他们对硬科学的参与仍旧非常有限。克罗斯比希望从生物学、医学、流行病学中获得能量，使他能对彻底的历史变革做出有力论证。事实上，他已经为历史研究带来了彻底变革。

1　艾尔弗雷德·W.克罗斯比：《哥伦布大交换：1492 年以后的生物影响和文化冲击》，康涅狄格州韦斯特波特：普雷格出版社，2003 年，第 xxvi 页。

第二部分：学术思想

5 思想主脉

要点 ✎━┓

- 艾尔弗雷德·W.克罗斯比《哥伦布大交换》的主要议题是病原体（引起疾病的微生物）和粮食作物的全球传播。

- 克罗斯比的主要论点是，欧洲疾病传播到新世界，以及新世界的农作物进入欧洲，为1492年克里斯托弗·哥伦布航行到美洲之后欧洲几个世纪的全球霸权铺平了道路。

- 克罗斯比围绕特定病原体、植物和动物的传播构建出了《哥伦布大交换》。

核心主题

在艾尔弗雷德·W.克罗斯比的《哥伦布大交换：1492年以后的生物影响和文化冲击》中，两个关键主题是：粮食作物和流行疾病的全球传播。其中克罗斯比主要考察了三个不同因素：

- 疾病在旧世界和新世界之间的迁移
- 新世界作物对旧世界人口的影响
- 两个半球之间动物和非粮食作物（如棉花和烟草）交换的结果

《哥伦布大交换》以革命性眼光，审视了流行疾病对美洲土著人口的影响。克罗斯比认为，天花等疾病的毁灭性影响，能帮助解释欧洲人如何主宰新大陆。他从作为拉丁美洲研究学者的经历中汲取营养，重新审视西班牙征服者对中、南美洲的迅速征服。他解释说："长期以来，我们只关注征服者的胆略，忽视了他们的生物盟友的重要性。"[1]

讨论过微生物对哥伦布大交换的影响之后，克罗斯比还研究了东、西半球间植物交换和动物交换的作用。他坚称，1492年之后，美洲粮食作物的传播，导致了旧世界人口的激增。[2] 克罗斯比将这两个关键主题相结合，勾勒出具体路径，展示了哥伦布从欧洲航海到美洲后一段时期，非人为因素决定性地影响世界历史的具体途径。

> "天花的肆虐对阿兹特克人和印加帝国造成的影响，很容易被20世纪的读者所低估。长期以来，我们只关注征服者的胆略，忽视了他们的生物盟友的重要性。"
>
> ——艾尔弗雷德·W. 克罗斯比，《哥伦布大交换》

思想探究

克罗斯比追溯了由细菌引起的疾病的危害，并得出令人信服的观点，这些疾病对新世界社会和政治的影响，超过了外部入侵或武力冲突。旧世界疾病涌入新世界，不仅直接造成死亡人数增加，还带来间接破坏，引发当地的社会和政治崩溃。克罗斯比提到，例如印加人 *，他们并未向西班牙人的军事力量屈服，却不得不屈服于西班牙人带来的疾病。几乎可以肯定，疾病造成了印加帝国领袖怀纳·卡帕克的突然死亡，这让民众在反抗侵略军时变得混乱而不堪一击。克罗斯比将这一点与他的更大主题相联系，认为人类历史由全球生态力量塑造而成。

在随后一章中，克罗斯比重申这一主题，认为性传播疾病梅毒从新世界向欧洲的传播，带来了巨大破坏。尽管不像天花在美洲的蔓延那样具有灾难性，梅毒在欧洲的传播也影响广泛，克罗斯比在其中也梳理出许多与梅毒相关的文化轶事。尽管关于梅毒迁移的最近研究显示，克罗斯比宣称的梅毒源自新世界的观点并不正确，但这个如今颇有争议的观点并未破坏克罗斯比认为疾病通过哥伦布大交换进行传播的中心议题。

克罗斯比还认为，美洲粮食作物的传播，导致了1492年后旧世界的人口大爆炸。[3]为证实他的观点，克罗斯比绕过许多历史学家在政治、冲突、殖民化等领域的研究，只关心从美洲流向欧洲的食物。玉蜀黍（玉米）、马铃薯和红薯（番薯）等美洲作物，与欧洲常见的粮食作物比起来，卡路里含量高得多。因此，这些食物后来成为欧洲人和亚洲人饮食中的重要组成部分。另一方面，牲畜输入到美洲，使西半球能够养活快速增长的欧洲移民人口。由于新世界农作物的传播，欧洲人口越来越多，成为移民到美洲的变革力量。克罗斯比指出，如果没有早些时候新世界粮食作物进入欧洲，就不会有后来旧世界人民迁往新世界。

语言表述

为强调生态交换的历史重要性，除粮食作物迁移外，克罗斯比还围绕疾病、流行病等主题提出更大论点。他让我们相信，非人为因素在哥伦布大交换中扮演着中心角色。他列举的一系列证据显示，是植物、动物或疾病带来了巨大改变，而这些改变曾经都被归因于技术变革，政治优势或文化力量。克罗斯比对早先发表过的两篇论文进行整合、扩展，构成了他论证的框架。

在《哥伦布大交换》中，克罗斯比提出了传统编史学*中存在的问题，主要关于如何在历史研究领域运用创新方法。尽管如今《哥伦布大交换》被公认为环境史著作，但在该书首次出版时，这还是个身份不明的领域，甚至该学科还未出现。那些从前依靠政治或经济术语来阐释的讨论，有赖克罗斯比的参与和重构，才形成了现有的学科。如此一来，《哥伦布大交换》以新的答案回应了熟悉的旧问题，比如"欧洲人如何征服新世界"，"哪些因素造成1492年后旧世界经济和人口的快速增长"。

《哥伦布大交换》还挑战历史学家，让他们认识到科学与历史的直接相关性，迫使他们涉足陌生领域，去触碰和发现考古学、生物学和流行病学中的新思潮。

1　艾尔弗雷德·W.克罗斯比：《哥伦布大交换：1492年以后的生物影响和文化冲击》，康涅狄格州韦斯特波特：普雷格出版社，2003年，第52页。

2　克罗斯比：《哥伦布大交换》，第52页。

3　克罗斯比：《哥伦布大交换》，第165—207页。

6 思想支脉

要点 ⚷—

- 艾尔弗雷德·W.克罗斯比研究了克里斯托弗·哥伦布1492年航行到新世界后的智性影响，并对欧洲人入侵带来的环境后果持悲观态度。
- 为探讨哥伦布大交换带来的智性影响，克罗斯比为后辈历史学家提供了一个模型，去研究思想和物质世界的相互作用。
- 自《哥伦布大交换》出版以来，历史学家对哥伦布航行到美洲的消极后果有了更深理解。

其他思想

在《哥伦布大交换：1492年以后的生物影响和文化冲击》中，克罗斯比还提出了一个重要的从属观点，关于哥伦布大交换如何在观念上影响大西洋两岸的智性体系。在该书第一章"对比"中，克罗斯比解释了对新世界人群、粮食作物和动物的发现是如何与欧洲人和美洲土著人对他们所在世界的基本信仰相一致的。例如，16世纪西班牙宗教思想家何塞·德·阿科斯塔*等认为，许多作品都能够解释南美土著居民的起源。这一观点非常有趣，因为它暗示了自然世界和观念世界的联系。但它仅是被提出，并未得到完全发展。

尽管人们对克罗斯比《哥伦布大交换》中弥漫的悲观色彩仍有争议，但书中展现的另一个重要的从属主题，主宰了书的结尾部分。克罗斯比在书末告诉我们，"哥伦布大交换留给我们的，不是更丰富而是更匮乏的基因库。我们星球上现有的生命种类，都比哥伦布时期要少，匮乏将愈加严重。"[1] 克罗斯比认为，人、植物、动物和微生物在两个半球间的迁移造成基因变化，导致生物同质性，人类和生物的多样性都受到压制。

> "哥伦布的到来，给美洲印第安人带来了灾难，但也留下一些令人愉快的后遗症，比如美洲草原上骑马文化的传播。但整体来说，白人和黑人的到来，带来了疾病，以及恐吓、驱逐、酗酒……还有对许多人的生命和他们的生活方式的抹杀。"
>
> ——艾尔弗雷德·W. 克罗斯比，
> 《哥伦布航海，哥伦布大交换和历史学家们》

思想探究

克罗斯比的悲观评估不限于结论部分；他贯穿全书都在阐释这个重要的副主题。他断言，"我们星球上现有的生命种类，都比哥伦布时期要少"，他有意识地引导我们，关注从前被忽略的探险时代生物和人口的损失。克罗斯比发现，生物同质性的增加和全球生物多样性*（全球生物种类的多样性）的减少，反映出1492年哥伦布从欧洲航行到美洲后，发生的关键性全球变化。人类文化受疾病威胁，甚至完全被其毁灭。某些植物和动物的数量缩减、消失或增加，完全取决于它们在新的全球经济中扮演的角色。

该书中的悲观色彩具有一定争议性，它与哥伦布作为一个历史大英雄的传统认知截然相反。然而，对克罗斯比来说，他的主要观点在于，强调哥伦布大交换中的非人为因素。美洲印第安人口骤减，为欧洲文明的全球统治铺平了道路；他们屈服于某些无法自然免疫的疾病，这削弱了他们抵御欧洲殖民者的实力。如克罗斯比所言，"陌生人群相遇，更多是他们之间细菌群的差异，而不是习俗的差异，造就了历史。"[2]

被忽视之处

尽管《哥伦布大交换》中的大部分观点都被人仔细研究过，但书中仍存在未受重视的部分。书中被严重忽略的或许是第一章"对比"，其中克罗斯比讨论了东、西两个半球间的差异。在讨论中，他审视了"美洲的问题"，也就是说，他探索了克里斯托弗·哥伦布"发现"美洲大陆对欧洲想象力的影响。[3] 尽管令人着迷，文本的这一方面仍有待开发。该书不是彻底周密的分析，而是对欧洲想象力的一系列有趣但不连贯的观察。克罗斯比写道，"美洲是一颗方钉子，要契入创世纪的圆孔里。"[4]

在书中，克罗斯比探讨了新世界如何改变欧洲人看待自己和环境的观念，他的这一举措被认为超越了时代。他认为，环境史学家过多关注物质现实，忽视环境观念，他本人在第一章的简短讨论中涉及了思想世界。克罗斯比思考的是，美洲的"发现"如何向基督教的创世纪神话发问，促使一些非正统的宗教思想家得出多次创造说的结论。克罗斯比并不是强调环境本身，而是在开篇第一章就显露出对环境问题的体察，并给出思考此类问题的方法。这一令人瞩目的观点使我们认识到，新世界的物种和人群如何极大地影响了包括基督教思想在内的欧洲的智识和精神结构。

1 艾尔弗雷德·W. 克罗斯比：《哥伦布大交换：1492 年以后的生物影响和文化冲击》，康涅狄格州韦斯特波特：普雷格出版社，2003 年，第 219 页。

2 艾尔弗雷德·W. 克罗斯比：《哥伦布航海，哥伦布大交换和历史学家们》，华盛顿特区：美国历史协会，1987 年，第 24 页。

3 克罗斯比：《哥伦布大交换》，第 12 页。

4 克罗斯比：《哥伦布大交换》，第 12 页。

7 历史成就

要点 🗝

- "哥伦布大交换"成为历史学家和社会科学领域学者的通用词汇。

- 《哥伦布大交换》出版时，正值环境史学家试图研究人类与自然环境关系的时期。

- 20世纪70年代以来，经济学家尝试通过定量指标——即通过包括统计在内的计量方法——来考查哥伦布大交换的影响。

观点评价

艾尔弗雷德·W. 克罗斯比1972年的著作《哥伦布大交换：1492年以后的生物影响和文化冲击》仍旧与学术界息息相关，但却略显过时，因为其中的论证已被学术界广泛接受。曾经革命性的方法如今已成共识，环境史学家探索人与环境的关系时，迫切想要融入科学研究和生态学。在讨论欧洲征服过程中流行疾病的确切作用时，如今的史学家和学者们通常都接受克罗斯比的观点，认为它扮演了中心角色。与此类似，学者们通常还认可克罗斯比的另一观点，即新世界食物对旧世界人口的重要性。关于这一讨论，经济学家试图进行更精确的量化研究。[1] 因此当前形成的共识是，克罗斯比在《哥伦布大交换》中的研究基本正确，还可进一步推动。

与此同时，《哥伦布大交换》始终被看作环境史领域的先锋之作。克罗斯比的方法非常成功，"哥伦布大交换"的思想如今被广泛接受，它合理解释了1492年克里斯托弗·哥伦布从欧洲航行到美洲后的变化。事实上，学者们都在自由使用这个词，不再从克罗斯比处引用。由于接受度太高，他的许多具体论证已经丧失了某些相关性，但他使用过的方法依然重要而切题。环境史学家 J. R. 麦克尼尔＊在 2003 版《哥伦布大交换》前言中称，克罗斯比的珍贵之处"不在于对哥伦布大交换记录的详尽性，而在于他建立了一种视角，一个理解生态和社会事件的模型"。[2]

> "克罗斯比的珍贵之处，不在于对哥伦布大交换记录的详尽性，而在于他建立了一种视角，一个理解生态和社会事件的模型。"
>
> ——J. R. 麦克尼尔，2003 版《哥伦布大交换》前言

当时的成就

1972 年《哥伦布大交换》出版时，人们还无法估量克罗斯比的论证和方法的新颖性。因为在当时，他的观点、主题和方法将该书置于主流历史研究之外。他费尽气力才找到出版社，该书最终付印后，一度被认为更应该是人类学或医学史著作。然而这部书本身，以及后来它在历史学界的广泛传播，都显示出如今克罗斯比的思想进入了史学领域的核心位置。这样的地位本应在该书初出版时就引发学者兴趣。

在《哥伦布大交换》出版前，流行的历史书写常受文化和政治假设所限，该书对这一传统方式提出挑战。为批评这些假设，克罗斯比指出，在驱动人类历史中起关键作用的是普世的生物问题。这一方法对欧洲中心主义＊叙事去中心化。这种叙事认为，欧洲文明的全球霸权和它在政治、文化、智识和技术上的优越性相关联。克罗斯比则认为，欧洲之所以能实现美洲殖民，以及随后几个世纪对

世界其他地区的殖民，不过是历史的偶然和生物因素汇聚的结果。

局限性

尽管《哥伦布大交换》在历史学科之外影响力有限，它还是产生了两个重要结果：第一，如我们所见，关于1492年克里斯托弗·哥伦布的航海，哥伦布大交换如今是人们普遍接受的思考框架。第二，克罗斯比在历史领域强调非人为因素，启迪了人文社会学科采用类似的方法。经济学、地理学等其他相关领域的学者都使用"哥伦布大交换"一词，却从不直接提及克罗斯比的作品，可见该观点的流行之广。

尽管克罗斯比的作品直接从各门学科中汲取营养，并受益于它们的影响，但要进行反向追溯并非易事，比如它如何影响了生物学和流行病学等领域。从某种意义上说，这些领域已开始追问此类问题，而克罗斯比一直期望将这些问题置于历史研究的中心位置。许多科学学科，尤其是公共健康等领域，都在采用更为历史化的方法，这似乎显示出克罗斯比著作的潜在影响力。

在除历史之外的其他社会科学中，《哥伦布大交换》也产生了显著影响。最近的经济学著作试图量化分析哥伦布大交换在人口增长中的重要性和影响。在发表于《经济展望杂志》的论文中，经济学教授南森·纳恩*和南希·钱*指出，在经济学文献中，利用哥伦布大交换研究生物影响常常被"忽视"，他们因此做了这方面的研究。[3] 研究结果很大程度上证实了克罗斯比书中的观点。他们发现，蔗糖、咖啡等最终成为重要食品的产品，它们在新、旧世界的不同价格间有着重要联系。也有类似研究成果，试图量化分析新世界流行疾病的蔓延程度。此类成果很大程度上都试图为克罗斯比的抽象论证增加定量论据。事实上，学界对克罗斯比某些思想的评估也有变化，如关于他对梅毒的起源和传播的说法。但迄今为止，克罗斯比的多数观点已被证实，并对历史以外的其他领域施加影响。

---

1 南森·纳恩、南希·钱："哥伦布大交换：疾病、食物和观念的历史"，《经济展望杂志》24 卷，2010 年第 2 期，第 163—188 页。

2 J.R. 麦克尼尔：《哥伦布大交换：1492 年以后的生物影响和文化冲击》前言，康涅狄格州韦斯特波特：普雷格出版社，2003 年，第 xiii 页。

3 纳恩、南希："哥伦布大交换"，第 164 页。

8 著作地位

要点 ⚷——

- 克罗斯比著作的整体目标是研究非人类行动者（尤其是植物、动物和疾病）如何塑造历史事件的进程。

- 艾尔弗雷德·W.克罗斯比的《哥伦布大交换》为他之后的工作制定了议程。

- 《哥伦布大交换》的出版确立了克罗斯比作为环境史学科奠基人之一的地位。

定位

艾尔弗雷德·W.克罗斯比的博士论文"美国、俄国、大麻和拿破仑"（1965 年成书出版），探讨了与《哥伦布大交换：1492 年以后的生物影响和文化冲击》截然不同的主题。前者主要研究美国独立战争 * 和 1812 年战争（都是年轻的美国为获政治和经济独立与英国间的武装冲突）期间，俄国与美国之间的（尤其有关大麻的）贸易策略。虽然克罗斯比的博士论文正是几年后他所摒弃的那类政治史，对他的著作的观察仍可发现：他的研究主题宏大，常追溯相去甚远的地域和政治事件间的意外联系。这之后又过了七年，到了 20 世纪 70 年代，克罗斯比才真正进入到环境史。

《哥伦布大交换》的灵感来自他之前发表的两篇论文，文章的某些部分被用进了书中。如关于流行病和西班牙征服的第二章，之前发表在名为《拉丁美洲历史评论》的期刊上。有关梅毒的第四章的某些部分，曾发表在《美国人类学家》上。这两篇文章和书中的相应章节基本相同，这也是为什么该书结构有些松散的原因。六篇相对独立的文章，组合成了克罗斯比的主要论证。从探索 1492 年之后的疾病开始，克罗斯比充实了《哥伦布大交换》中的研究，他试图追溯哥伦布航海的全部影响，让读者更好地理解其背后的生物学原因。

> "20 年前，我完成了这本书，关于哥伦布航海如何影响世界人民……我千方百计想联系一家出版社，正当快要放弃时，格林伍德出版社主动写信问我，手头是否有东西要出版……这本书的销售稳定而持久，差不多每年三千本，到现在总共卖了四、五万本。"
>
> ——艾尔弗雷德·W. 克罗斯比，"重估 1942"

整合

《哥伦布大交换》是克罗斯比第一本重要的图书出版物，该书为他整个职业生涯的研究问题确立了纲领，他未来的著作都在研究人类如何在更广泛的生态关系之网中生存。《哥伦布大交换》之后，克罗斯比持续追问跨越若干世纪的全球变化。《生态帝国主义：900—1900 年欧洲的生物扩张》一书，论述了横跨一千年的人类历史。[1] 在这部 1986 年出版的开创性著作中，克罗斯比重申了他在《哥伦布大交换》中首次提出的观点，认为是生物因素而非军事力量促成了欧洲帝国的成功。该著作研究的时间范围大大扩展，克罗斯比回溯到古挪威人入侵和十字军东征，考察 19 世纪太平洋地区"新欧洲"的形成。在 1987 年的短篇《哥伦布航海、哥伦布大交换和历史学家们》中，他对哥伦布航海的编史学问题做了严格审视。

克罗斯比的著作主题统一，它们大量分析了两个半球间人、植物、动物、食物和疾病的交换。然而，多年来克罗斯比不断扩展研究的地域范围。在《哥伦布大交换》中，他主要研究美洲。在《生态帝国主义》中，他将欧洲的疾病、动物和农业的扩张置于全球欧洲帝国主义的中心。他在《哥伦布大交换》中对疾病的兴趣，延续到他的后期作品《被遗忘的传染病》[2]中，这部颇具影响的历史著作是关于 1918 年大流感 * 的（流感大规模爆发，全球近五亿人染病，造成五千万到一亿人的死亡）。[3]

意义

《哥伦布大交换》极具影响，根据随后的历史文献来看，克罗斯比论证的独特性和原创性难以估量。他关于历史中非人为因素的重要性的许多主张，都已成为主流历史学家的基本假设。

该作品是环境史学科奠基著作之一，关于环境史学家如何从科学以及地理学、人类学等学科中汲取养分，重新诠释关键的历史时刻，它也提供了最强大的例证。

《哥伦布大交换》的中心观点对具体的环境史学和整个历史研究都有影响。例如，流行病是欧洲征服新世界的决定性因素，这一观点极为重要。早期的学者多认为，欧洲的技术以及美洲土著居民在宗教和政治体制上的"原始性"是欧洲征服的根本原因。但克罗斯比通过展示疾病的影响力，推翻了前人对该段历史的书写，强化了非人为因素的重要作用。类似地，克罗斯比对"哥伦布大交换"的强调，引发了一种支配性的历史范式，或者说概念模式，它以与此前出版物完全不同的方式，诠释了一段重要的历史时期。

　　这些观点首次出现带来的影响，以及未来一些年的后续影响，都确保了克罗斯比的作品始终处于一系列伟大的历史著作之列。

1　艾尔弗雷德·W. 克罗斯比：《生态帝国主义：900—1900 年欧洲的生物扩张》（第二版），剑桥：剑桥大学出版社，2004 年。

2　1976 年在康涅狄格州格林伍德出版社首次出版，当时名为《流行病与和平》。

3　艾尔弗雷德·W. 克罗斯比：《被遗忘的传染病：美国 1918 年的流感》（第二版），剑桥：剑桥大学出版社，2003 年。

第三部分：学术影响

9 最初反响

要点 🔑

- 艾尔弗雷德·W.克罗斯比认为，生物因素是西方文明统治的主因，《哥伦布大交换》的早期评论者对此持反对意见。

- 作为对自己著作的经验批评（即基于可见证据得出的批评）的回应，克罗斯比承认梅毒不是从美洲带入欧洲的。

- 克罗斯比的著作得到新一代历史学家的支持，他们为欧洲文明的全球霸权找到了不同于以往的解释。

批评

艾尔弗雷德·W.克罗斯比《哥伦布大交换：1492年以后的生物影响和文化冲击》的早期批评大部分针对的是书中最后一个章节中深刻的悲观主义。"我们这个星球上现有的生命种类，"克罗斯比观察道，"都比哥伦布时代要少。"[1]

克罗斯比认为，1492年克里斯托弗·哥伦布从欧洲航行到美洲之后，关键性的全球变化是生物同质性的增加和全球生物多样性的减少共存。许多评论者抓住这一点，认为这种观点夸张且富有争议。地理学者兼加州大学洛杉矶分校荣誉教授加里·S.邓巴 * 认为，该书的"结论过于悲观"。[2] 正是在结论处，人们发现了克罗斯比的著作引发争议的最初线索：他认为，视欧洲为全球主导的传统叙事过分乐观，忽视了欧洲霸权的残酷性。美国有（现在仍有）以哥伦布命名的全国性节日，而克罗斯比却在告诉我们，哥伦布航海让全球每个人过得更糟了。

对克罗斯比的悲观主义的早期批评预示了未来辩论所采取的立场。克罗斯比和地理学者兼畅销书作家贾雷德·戴蒙德*，都试图将欧洲占主导地位的情况视为非人为因素的结果。许多学者则回应称，文化对经济和政治成功起着决定性作用。美国经济史学家大卫·兰德斯挑战克罗斯比的观点，认为政治、社会和文化因素才是成败的关键性决定因素。[3]《哥伦布大交换》的早期接受暗示了它未来遭遇的争议，说它一出版就广受关注是不准确的。在 20 世纪 70 年代初，学者们还未意识到克罗斯比的论证的巨大力量。

> "30 年前，克罗斯比的观点遭多数历史学家冷漠以待，被许多出版商拒绝，受至少部分评论者敌视。而如今，在对现代历史的描绘中，他的观点占据显著位置。"
>
> ——J. R. 麦克尼尔，2003 版《哥伦布大交换》前言

回应

克罗斯比对《哥伦布大交换》的批评声音的回应，可见于他随后的著作《生态帝国主义》（1986）中，以及 2003 版《哥伦布大交换》的序言中。克罗斯比承认，依照最新的科研成果，他误判了梅毒的起源，最终他修正了自己的结论。从某种意义上说，这一批评也反映了他的著作的特别之处；由于他对科学研究的涉足非常深入，他的某些观点就会引起注意或遭到反驳。在 2003 版序言中，克罗斯比还承认，关于流行病如何为欧洲殖民统治铺平道路，他的描述缺乏足够的细节去涵括非生物因素。事实上，早在 1976 年，在"处女地流行病作为美洲土著人口下降的原因"一文中，他对此给出了细致的解释。[4]

然而在其他领域，克罗斯比坚守自己的阵地。他始终反对将欧洲优越性更多归于文化和政治结构而不是非人为因素。事实上，他在《生态帝国主义》中进一步抨击了这一观点。克罗斯比追溯了新世界里他称作"新欧洲"的发展，断言生物和生态进程处于帝国主义的中心。1492 年哥伦布航海之后，对于以殖民主义为标志的欧洲对世界的主宰，各种解释至今仍无定论。但在历史学专业领域内，克罗斯比的阐释已被广泛接受。

冲突与共识

克罗斯比的一些实证结论被修正或遭到反驳，他关于哥伦布大交换把梅毒带到欧洲的推论，就是其中之一。如今的证据表明，1492 年以前，这一疾病在欧洲已经存在。但他作品中比较重要的观点仍令人信服：他通过结合人类学、地理学、生物学及其他领域的论据指出，由哥伦布大交换带来的非人为因素，是变化的决定性力量。

最近的一些研究成果，一方面承认非人为因素（在克罗斯比看来，包括植物、动物、疾病）的重要性；另一方面认为它们不应与人类文化完全割裂开。[5] 这类文献模糊了自然和文化的界线，认为它们从根本上就是相互交织的。在关于流行病和新世界征服的章节中，克罗斯比对此观点也有所暗示，但未直接提及。在"处女地流行病作为美洲土著人口下降的原因"一文中，克罗斯比清晰地阐明，流行病的影响并非完全是生物性的，而应将人为和非人为因素综合起来考虑。[6]

在克罗斯比整个的研究成果中，他的后期著作是对前期观点的发展，而不仅仅是对关于《哥伦布大交换》的批评做出的回应。事实上，克罗斯比在作品中做了建设性批评，对关于《哥伦布大交换》的评估做出回应，对认为欧洲的统治地位主要受欧洲文化支配的观点，他仍旧进行了反驳。

1　艾尔弗雷德·W.克罗斯比：《哥伦布大交换：1492 年以后的生物影响和文化冲击》，康涅狄格州韦斯特波特：普雷格出版社，2003 年，第 219 页。

2　G. S. 邓巴："哥伦布大交换评论"，《威廉与玛丽季刊》30 卷，1973 年第 3 期，第 543 页。

3　大卫·S.兰德斯：《国富国穷：为什么有的国家富有的国家穷》，纽约：诺顿，1998 年。

4　艾尔弗雷德·W.克罗斯比："处女地流行病作为美洲土著人口下降的原因"，《威廉与玛丽季刊》33 卷，1976 年第 2 期，第 289—299 页。

5　参见威廉·克罗农：《土地的变迁：新英格兰的印第安人、殖民者和生态》，纽约：希尔和王出版社，1983 年；理查德·怀特：《依赖之根：乔克托族、波尼族和纳瓦霍族之间的生存、环境和社会变化》，林肯：内布拉斯加大学出版社，1983 年。

6　克罗斯比："处女地流行病"，第 289—299 页。

10 后续争议

应用与问题

艾尔弗雷德·W.克罗斯比之后的学者通过重要的方式拓展了他的著作。《哥伦布大交换：1492年以后的生物影响和文化冲击》是一个必要的纠正。在它出版的时代，历史书写主要围绕政治或社会因素来进行。一些新近的著作，试图解释人为和非人为因素间的互动关系，这种方法克罗斯比在《哥伦布大交换》中就有所暗示，但在后期的著作中才明确涉及。事实上，克罗斯比承认，他对非人为因素的强调，开启了当今的学术研究，这种研究试图打破人类和非人类间的界限。例如，这部作品认为，天花在欧洲征服中起着决定性作用。但值得注意的是，天花病毒的威力，和社会、政治结构共同作用，才造成这种流行病的致命危害。因此，这些新近的著作更有力地宣称，所有流行病，都是生物和文化因素共同作用的结果。

依循这样的思路，关于历史框架中生物和人类如何相互依存，学者们有了更细致的理解。克罗斯比期望《哥伦布大交换》成为某种补救措施，他试图说服历史学家，让他们放弃从前珍视的政治和社会观念，转而思考疾病和新型农作物的重要性。尽管他成功实现了目标，但后辈历史学家们不得不发展出更强大的理论框架作为参考。这些方法消解了自然和文化间的任意性区分，质疑克罗斯比提出的人为和非人为因素能清晰分开的假设。然而，像这种理论上更微妙和统一的叙述，比如美国环境史学家理查德·怀特的《有机机器》(1995)，在克罗斯比的基础性著作出版之前，很可能是难以出现的。

> "凭借丰富的历史实验，《哥伦布大交换》让经济学家对历史作用于经济的长期影响产生兴趣。因此，经济研究开始重点关注，如何通过殖民主义将欧洲体制移植到了非欧洲的世界其他地区。"
>
> ——南森·纳恩、南希·钱，
> "哥伦布大交换：疾病、食物和观念的历史"

思想流派

《哥伦布大交换》因帮助历史学家更新研究方法而闻名，鼓励"追随者"倒在其次。在 20 世纪 70 年代末到 20 世纪 80 年代之间，历史学家们将环境史纳入了主流领域。其中帕特里夏·利默里克*、J. R. 麦克尼尔、理查德·怀特、唐纳德·沃斯特*等人，从严格意义上说，都不是克罗斯比的信徒。他们都涉及了克罗斯比的主题，但也从其他年长的学者，比如美国环境史学家沃尔特·普雷斯科特·韦伯那里，汲取了养分。从某种意义上说，在迈向环境史的进程中，《哥伦布大交换》的确是一个关键性文本。

克罗斯比以生态视角研究历史，这种特殊方法的追随者包括怀特和颇有影响的环境史学家威廉·克罗农。像克罗斯比的《哥伦布大交换》一样，怀特的《有机机器》[1]和克罗农的《土地的变迁》（1983）[2]所关注的，是对与人类和社会相交织的植物、动物、微生物之间的生态关系的历史考量。美国环保主义者奥尔多·利奥波德[*]最先提出借用生态方法书写历史，克罗斯比作为专业历史学家，第一个接受了这一挑战。他不仅影响了生态史，在鼓励历史学家运用科学文献方面，也发挥着决定性作用。

当代研究

对于学者们理解 1492 年克里斯托弗·哥伦布航海的结果，克罗斯比的《哥伦布大交换》起到了决定性的重塑作用。他的书有力地证明，环境史对所有历史研究领域都很重要。该书给环境史学家带来了深远影响。如今的历史学者们广泛涉足地质学和流行病学等领域，哥伦布大交换已成为公认的历史现象。历史学家不再质疑大交换是否发生，而是尝试评估和确定它的范围。最新的历史著作已试图对生物因素和社会实践的相互作用加以理论化，经济学家们则对交换的程度进行了量化研究。[3]

如今的环境史学家，尽管对《哥伦布大交换》的某些细节略有异议，但还是普遍接受了它的中心假设。然而，说该著作创建了一个统一的流派或方法并不准确。它没有试图开发一批克罗斯比的拥护者；而是示范了一种强调生态和环境问题的、检视历史的具体方式。《哥伦布大交换》一出版，这种方式就迅速成形了。

受该书启发的学者中，几乎没有人认为自己是克罗斯比的信徒。但环境史作为一门学科，从克罗斯比处获益很大。使这门学科获得重大发展的，还包括略早时期的一些学者和思想家，比如沃尔特·普雷斯科特·韦伯、弗雷德里克·杰克逊·特纳和奥尔多·利奥波德。

1 理查德·怀特:《有机机器：哥伦比亚河的改造》，纽约：希尔和王出版社，1995年。

2 威廉·克罗农:《土地的变迁：新英格兰的印第安人、殖民者和生态》，纽约：希尔和王出版社，1983年。

3 南森·纳恩、南希·钱:"哥伦布大交换：疾病、食物和观念的历史"，《经济展望杂志》24卷，2010年第2期，第163—188页。

11 当代印迹

要点 ⚷

- 艾尔弗雷德·W. 克罗斯比的《哥伦布大交换》被视为环境史领域的先锋著作。

- 传统的历史叙事认为，1492 年后欧洲的统治地位依赖文化、政治、社会因素，克罗斯比在该书中对此观点提出挑战。

- 克罗斯比的反对者强调，在解释近代早期（约 15 世纪末到 18 世纪末）欧洲的成功和实力时，文化因素更加重要。

地位

在 1972 年的著作《哥伦布大交换：1492 年以后的生物影响和文化冲击》中，艾尔弗雷德·W. 克罗斯比提出的专业问题在学术界仍是争议性话题。尽管学者们普遍接受了他的观点，但仍有一些问题，如新世界食物的具体影响或旧世界疾病的蔓延的重要性，引发了深入的探究和有针对性的批评。如今，哥伦布大交换的概念成为公认的框架，用来解释过去 500 年重要的历史特征。克罗斯比在解释欧洲的实力和它对西半球的征服时，强调非人为因素的重要性，这引起一些著名的学术评论家的关注。

克罗斯比的观点在 20 世纪 70 年代初刚兴起时，一度被认为是革命性的，因为它打破了传统范式（即解释模型），但某些新近著作对克罗斯比论点的理论基础提出了挑战。据观察，克罗斯比似乎认为，历史学家可以将文化和自然分开考虑；而事实上，它们是相互联系的。此外，关于哥伦布大交换对于欧洲和美国智性历史的重要性，克罗斯比的讨论被认为并不完善，一些新著作更有力地强调了思想和环境的关系。在 21 世纪环境史的语境中，《哥伦布大交换》因其自身的成功变成了受害者。它在历史研究中被运用得如此彻底，曾经革命性的观点似乎变成了陈词滥调，缺乏新近著作中的细微精妙之处。

> "《哥伦布大交换》刚出版时，克罗斯比通过生物学研究历史的方法很新潮。而如今，这本书被认为是环境史领域的奠基著作了。"
>
> ——梅根·甘比诺，
> "艾尔弗雷德·W. 克罗斯比的哥伦布大交换"

互动

《哥伦布大交换》的许多重要观点如今已被广泛接受，在历史专业范围内，克罗斯比几乎不再遭遇什么挑战，部分原因是他对环境史学家的影响非常大。但他思想中的某些方面，在学术界的其他一些领域仍旧不受欢迎。具体而言，这些反对意见关注克罗斯比论点中的暗示，认为这是对欧洲全球统治的文化阐释进行攻击。对于克罗斯比立场的怀疑来自保守派学者，尤其是经济学界的学者。他们优先考虑欧洲政府和金融机构的发展，认为这些才是造成近代早期变革的原因。

《哥伦布大交换》也被批评为过于关注客观事物，因为它忽视了人类对环境的认知和态度的重要性。克罗斯比自己也意识到，像他这样的学者，"更倾向于对客观事物感兴趣，而不是对认知客观事物这一行为感兴趣。他们对正在处理的事物的真实性从不怀疑，也从不担心他们没有能力抓住真相"。[1]学界较新的观点认为，这是一个严重的疏忽，因为人类对环境和生态的认识，在历史上具有重要作用。

持续争议

此前的历史叙事认为，欧洲帝国主义的成功在于先进的技术、社会和政治，克罗斯比认为应将此原因去中心化，但遭到一些人的质疑。尽管多数学者赞同克罗斯比，但也有一部分，尤其是经济史领域的学者，对他持反对意见。如果对比美国地理学家杰瑞德·戴蒙德和美国经济史学家大卫·兰德斯的两部著作，我们就会发现两种截然不同的观点。在畅销作品《枪炮、病菌与钢铁》（1988）中，作者戴蒙德，这位克罗斯比的间接拥护者，提出的观点与《哥伦布大交换》中的观点非常相似。他认为，地理和自然世界非常有助于解释特定人群和国家的成功或失败。

然而，在《国富国穷》（1998）中，兰德斯认为，在解释近代早期欧洲的成功和实力时，文化因素才更加重要。尽管学术界支持克罗斯比和戴蒙德的一方力量更强，但经济学领域大量的定量研究（即利用统计学等可测量的论据进行的研究）还是支持兰德斯的观点。虽然不受历史学家欢迎，但在那些相信欧洲的成功源自西方文化价值的人当中，《国富国穷》赢得了广泛的支持。

克罗斯比的学术评论者，都在以对自己有利的方式据理力争，大部分公共对话都被严重政治化。克罗斯比自己解释说，他的这本著作，源自20世纪60年代的动荡和对保守派前辈们的幻灭。通常来说，认为欧洲的支配地位源自文化因素的人，是政治上的保守派，他们相信欧洲文明的天然优越性。如历史学家玛格丽特·雅各布＊评论称，关于西方的成功问题，"对左派来说已经不合时宜……从广义上说，它已经被托付给右派了"。[2] 结果是，关于这些事件的公共对话反映出明显的极端化。政治上的左派支持者，把强调文化因素视为欧洲中心主义（源自某种对欧洲优越性的假想），视为为殖民主义的暴行做辩护。与此同时，右派人士则认为，把欧洲的支配地位归于自然或偶然，这否认了西方民主国家所珍视的价值。

公共话语缺乏某种复杂性，尖锐的分歧将问题分为自然和文化两方面。但辩论的双方都应当承认，某种程度上，现实反映出的是二者的融合，若说有什么动机，那就是双方要互相倾听。

---

1 艾尔弗雷德・W. 克罗斯比："环境史的过去和现在"，《美国历史评论》100 卷，1995 年第 4 期，第 1188 页。

2 玛格丽特・雅各布："思考不合时宜的想法，提出不合时宜的问题"，《美国历史评论》105 卷，2000 年，第 495 页。

12 未来展望

要点 ⚷

- 艾尔弗雷德·W.克罗斯比的作品仍将是环境史领域的奠基著作。
- 对那些试图研究历史进程中非人为因素影响力的历史学家来说，克罗斯比的文本将发挥持续影响力。
- "哥伦布大交换"仍被用作构想某一历史时期的框架。

潜力

艾尔弗雷德·W.克罗斯比的《哥伦布大交换：1492年以后的生物影响和文化冲击》，如果非要被视为一部基础性文本的话，它在流通方面可能并不具有持续的相关性。克罗斯比提出的观点、所进行的论证，曾经也将继续给编史学以持久的影响力。如今，环境史获得了应有的地位，许多历史学家都在思考人类世界和非人类世界的关系。克罗斯比在这其中功不可没。

克罗斯比这部1972年的著作非常成功，它引发的许多对话已远远超出自身的学科边界。有关新世界食物对欧洲人口增长的影响，克罗斯比的论证略显抽象，也缺乏精确性，这使得越来越多的著作进入该领域，对这一影响进行量化研究。[1] 对于社会结构和流行病发展的关系，学者们也有过类似的讨论。即使研究已远远超出《哥伦布大交换》的范围，克罗斯比对每次争辩和讨论的影响仍清晰可见。

此外，尽管最终《哥伦布大交换》很可能被认为是一部具有历史重要性的著作，而不是一部"活生生"的文本，它仍然与研究相关。此前少有著作能够做到把诸多种文献综合进连贯的历史研究中。从这个意义上说，这本书对后来的编史学非常重要，作为一部经典的历史著作，它将拥有持久的影响力。

> "由于克罗斯比的作品获得认可，'哥伦布大交换'这个词如今被广泛使用，用来描述自哥伦布以来，复杂多面的生态交换和影响链。"
>
> ——路易斯·德·沃西，"欧洲相遇：发现与探索"

未来方向

学者们将继续研究克罗斯比思想的发展和起源。《哥伦布大交换》讨论过动物和植物在两个半球之间的迁移，早期美国史学者弗吉尼亚·德约翰·安德森＊在新书中明确指出，动物自己就能进行殖民拓展，它们是欧洲殖民扩张的强有力盟友。[2]安德森的著作关注殖民化的新英格兰，与此同时，动物在整个殖民史中及世界不同地区的重要作用将成为未来研究的一个必然方向。与此类似，新作物传播的影响也一直受到热烈关注，历史学家和经济学家都将继续对这种影响进行量化研究。

环境史学家 J. R. 麦克尼尔也将克罗斯比的见解推向了新的方向。在 2003 版《哥伦布大交换》前言中，麦克尼尔回忆起这本书如何影响了他作为一个历史学家的思想："第一次遇见这本书，是1982 年一个雨天的下午，在一间临时使用的办公室里，我从一个齐肩高的书架上取下它，一口气读完，连吃晚饭都忘记了。"麦克尼尔 2010 年的著作《蚊子帝国》，应用了克罗斯比在《哥伦布大交换》中形成的研究方法，得出生态方面的答案，来回答长期困扰研究加勒比地区的历史学家的一个问题。该问题是，在 17、18 世纪，为什么大西洋地区的其他力量没有能够把西班牙这股衰败的势力驱逐出加勒比海地区？

小结

在论证和实施方面，《哥伦布大交换》堪称过去 50 年间最重要的历史著作之一。克罗斯比令人信服地指出，1492 年克里斯托弗·哥伦布从欧洲航行到美洲带来的最重大影响都源于非人为因素。尽管如今这一观点是广为接受、显而易见的，但在 1972 年该书刚出版时，它却是革命性的。此外，《哥伦布大交换》仍旧无可比拟，它教给人们如何以跨学科的方式，从各门学科中获取养分，从事历史研究。

克罗斯比凭借精湛的技巧和创造力，深入当时的科学文献，展示出这些领域对历史学家的巨大教益。作为补充，他也鼓励历史学家把他们的想法贡献给流行病学、生物学以及其他科学。因此，作为环境史的奠基著作，克罗斯比的这本书值得继续关注，它也是真正的跨学科研究的最好范例。

事实证明，无论是具体论述，还是宏观概括，《哥伦布大交换》的中心思想都极具影响力。例如，当克罗斯比提出，欧洲征服新世界的决定性因素是流行病的传播时，这一观点颠覆了整个历史学界。这标志着对早期学者著作的完全背离，因为他们都更强调欧洲的技术和美洲土著居民"原始的"宗教和政治体系。

为证明疾病的影响，克罗斯比推翻了前人的论证，坚持认可疾病、农作物和动物等非人为因素的重要作用。同样地，他对"哥伦布大交换"这一研究框架的强调，为如今占主导地位的历史分期范式（即关于如何划分历史时期）奠定了基础。该书刚出版时这些观点带来的影响，以及随后几十年它们的后续影响，都使这本书居于伟大历史著作之列。就像后哥伦布时代，新世界农作物养活急速增长的欧洲人口一样，《哥伦布大交换》像一颗种子，它的成长滋养了一大批史学家、学者和思想家。

1　南森·纳恩、南希·钱："哥伦布大交换：疾病、食物和观念的历史"，《经济展望杂志》24卷，2010年第2期，第163—188页。

2　弗吉尼亚·德约翰·安德森：《帝国的动物：家畜如何改变早期的美洲》，牛津：牛津大学出版社，2006年。

术语表

1. **美国独立战争**：1775 年至 1783 年，由北美的 13 个英国殖民地发起的独立战争，对抗英国的统治，后形成美利坚合众国。

2. **阿兹特克人**：美洲原住民的一支，最早可追溯至公元前 1100 年左右，主要生活在墨西哥中部山谷。

3. **生物多样性**：对地球上已有的生物种类变化的衡量。

4. **生物同质性**：特定环境中生物物种趋于减少的过程。

5. **黑人权力运动**：美国的一项权力运动，于 20 世纪 60 年代达到高潮，目的在于提升非裔美国人的集体利益。

6. **民权运动**：20 世纪 50 年代到 20 世纪 60 年代美国的一项社会运动，旨在消除种族歧视，为非裔美国人赢得公民权利。

7. **殖民化**：离开原居住地，在另一块土地上定居并实施政治控制的一种行为。

8. **征服者**：专指来自西班牙帝国，征服新世界的探险家和军人。最著名的征服者有：埃尔南多·科尔特斯（1485—1547），阿兹特克人的征服者；弗朗西斯科·皮萨罗（1475—1541），印加帝国的征服者。

9. **人口统计学**：利用人口的出生、死亡和疾病的统计数据进行的统计研究。

10. **流行病学**：针对疾病的流行进行的动态研究。

11. **欧洲中心主义**：对欧美文明的文化独特性和优越性的信仰。

12. **编史学**：对史学家使用的方法的研究；历史作为一门学科的演变。

13. **整体性**：与整体研究相关的方法，同部分研究相反。

14. **帝国主义**：通过殖民化和军事力量扩大国家影响力的政策。

15. **印加文明**：一个以现在的秘鲁为中心的前哥伦布时代文明。

16. **1918 年大流感**：1918 年 1 月至 1920 年 12 月爆发的一场严重的流感，全世界近 5 亿人感染，造成 5 000 万至 1 亿人死亡。

17. **新世界**：起源于 16 世纪初的一个术语，指被欧洲探险者登陆并征服的西半球陆地。

18. **旧世界**：非洲、欧洲和亚洲大陆，在 15、16 世纪欧洲探险者发现美洲前就为人所知，是当时已知世界的总称。

19. **病原体**：一种传染性生物机体，给宿主带来疾病。

20. **可能论**：一种理论，认为环境限定但并不决定人类社会中社会、文化的发展。

21. **天花**：一种传染性疾病，其病毒包括大天花和小天花。

22. **西班牙征服**：始于 1492 年，西班牙王国对南美洲长达三个世纪的扩张过程。

23. **梅毒**：通过性接触传播的疾病，它的症状包括皮肤溃疡和大脑功能受损。

24. **越南战争**：1954 年至 1975 年的一场长期冲突，信仰共产主义的北越与南越政府的交战，南越的主要盟友是美国。

25. **1812 年战争**：1812 年到 1815 年美国和英国之间的冲突，部分原因是英国试图限制美国的贸易。

人名表

1. **何塞·德·阿科斯塔**（1539—1600），西班牙耶稣会传教士和博物学家，因在南美洲自然史方面做出的开拓性贡献而著名。

2. **弗吉尼亚·德约翰·安德森**（1947年生），科罗拉多大学博尔德分校早期美国史教授，她因2006年出版的作品《帝国的动物：家畜如何改变早期的美洲》而著名。

3. **雷切尔·卡森**（1907—1964），美国海洋生物学家兼环保主义者，她因1962年出版的《寂静的春天》而闻名，该书主要探索了农药使用造成的环境危害。

4. **皮埃尔·乔努**（1923—2009），一位专门研究拉美史的法国历史学家，他还研究16—18世纪法国的社会和宗教史。

5. **弗雷德里克·E.克莱门茨**（1874—1945），美国植物生态学家，植物进化研究的先驱。在其最著名的"顶极群落"理论中，他认为，生态扰动后，植被会自然地趋向一个成熟的顶极状态。

6. **克里斯托弗·哥伦布**（1451—1506），意大利探险家，在西班牙王室的支持下，他完成了四次穿越大西洋的航行，发起了西班牙对新世界的殖民。

7. **威廉·克罗农**（1954年生），威斯康星大学麦迪逊分校的历史、地理和环境研究教授。他的名作《自然的大都市：芝加哥和大西部》获1992年班克罗夫特奖。

8. **贾雷德·戴蒙德**（1937年生），地理学家，任教于加州大学洛杉矶分校，以其作品《枪炮、病菌和钢铁》（1997）而闻名。

9. **加里·S.邓巴**（1931—2015），加州大学洛杉矶分校的一名美国地理学家，他因19、20世纪地理学学科史的研究而闻名。

10. **查尔斯·埃尔顿**（1900—1991），英国动物学家兼动物生态学家。

11. **怀纳·卡帕克**（1464—1527），印加皇帝，16世纪早期西班牙人入侵南美洲后，染天花死亡。

12. 玛格丽特·雅各布（1943 年生），加州大学洛杉矶分校杰出的历史学教授，其新作关注工业革命的文化起源。

13. 大卫·兰德斯（1924—2013），美国经济史学家，因其工业革命方面的著作而闻名。

14. 奥尔多·利奥波德（1887—1948），美国环保主义者和荒野保护的倡导者。

15. 帕特里夏·利默里克（1951 年生），美国环境史学家，研究美国西部的杰出史学家之一。

16. J. R. 麦克尼尔（1954 年生），乔治敦大学环境史学家和历史学家，他的代表作是 2000 年出版的《阳光下的新鲜事：20 世纪世界环境史》。

17. 罗伯特·E. 穆迪（1901—1983），美国历史学家，职业生涯的大部分时间在波士顿大学工作，专攻早期美国史。

18. 南森·纳恩，加拿大人，哈佛大学经济学教授，他的著作关注经济史和发展经济学。

19. 南希·钱，耶鲁大学经济学副教授，研究领域为饥荒和经济发展史。

20. 弗雷德里克·杰克逊·特纳（1861—1932），20 世纪初美国历史学家，因撰写关于美国边疆的著作而闻名。

21. 查尔斯·韦林登（1907—1996），比利时籍中世纪历史学家，专门研究经济史和殖民主义史。

22. 保罗·维达尔·白兰士（1845—1918），法国地理学的奠基人之一，以"可能论"而闻名，该理论反对环境决定论（即认为环境不是人类社会形态的根本性决定因素）。

23. 沃尔特·普雷斯科特·韦伯（1888—1963），美国历史学家，在美国西部环境史方面有开创性贡献。

24. 理查德·怀特（1947 年生），斯坦福大学美国环境史学家，其代表作为 1983 年出版的《依赖之根：乔克托族、波尼族和纳瓦霍族间的生存、环境和社会变化》。

25. 唐纳德·沃斯特（1941 年生），堪萨斯大学著名的美国历史学教授，环境史的创始人之一，代表作是出版于 1979 年的《沙尘暴：20 世纪 30 年代的南部平原》。

WAYS IN TO THE TEXT

- Alfred W. Crosby is an American environmental historian and professor emeritus at the University of Texas at Austin.

- In *The Columbian Exchange*, Crosby argues that the movement of plants, animals, and diseases between Europe and the Americas were the most important legacy of the voyages of the fifteenth-century Italian explorer Christopher Columbus,* considered a key figure in the history of European colonization.

- *The Columbian Exchange* claims that ecological factors were important to the European colonization of the Americas.

Who Is Alfred W. Crosby?

Alfred W. Crosby, the author of *The Columbian Exchange: Biological and Cultural Consequences of 1492* (1972), was born in Boston, Massachusetts, in 1931. He graduated from Harvard University in 1952 before serving in the US Army from 1952 to 1955, stationed in Panama. After the army, he received his Masters in the Art of Teaching from the Harvard School of Education and a PhD in history from Boston University in 1961, working under the direction of Robert E. Moody,* a noted historian of American colonial history. Crosby's dissertation, *America, Russia, Hemp, and Napoleon*, was published in 1965 as his first book; it traced an economic history of the trade between the United States and the Baltic from 1783 to 1812.

Crosby, however, quickly became a critic of the mainstream conventional historical scholarship of his time. His work grew out of the political and intellectual developments of the period;

activism based on environmental concerns had been gathering steam since the American biologist Rachel Carson's* influential book *Silent Spring* (1962), which explored the ecological implications of pesticide use. Similarly, Crosby explains in his preface to the 2003 edition that *The Columbian Exchange* emerged from the social upheaval of the 1960s, sparked by America's Civil Rights* and Black Power* movements (key features of the struggle of black Americans for civil and political equality), as well as the Vietnam War* (a conflict in which the United States fought on behalf of South Vietnam against communist North Vietnam at the cost of many thousands of lives).

Previous generations of historians had focused on writing celebratory political histories. In his preface, Crosby observes that "for these men ... the good guys always won."[1] But the tumult of the 1960s raised doubts in Crosby's mind about these narratives and he began to consider approaches to history that considered matters beyond the political. This led him to studies in the areas of science and medical anthropology.

During his career, Crosby taught at Ohio State University and Yale. Although he has spent the bulk of his time at the University of Texas at Austin, he wrote and researched *The Columbian Exchange* while at Washington State University. As of 2015, Crosby is professor emeritus of history, geography and American studies at the University of Texas, Austin.

What Does *The Columbian Exchange* Say?

Earlier historians argued that in the period following Christopher

Columbus's 1492 voyage from Europe to the Americas (an expedition considered to have vital consequences for European colonialism) history was principally driven by social and political factors. In *The Columbian Exchange*, however, Crosby insists that biological factors were of particular significance to the changes that occurred. He outlines a "Columbian Exchange" in which previously isolated plants, animals, microbes, and peoples came into contact, resulting in a trend towards global biological homogeneity*—in other words, as some plants and animals spread worldwide, their competitors disappeared or were marginalized. In explaining the success of European colonization in the Americas, Crosby speaks of biological shifts, such as the importation of high-calorie foods to Europe, including potatoes, manioc (also known as cassava), and corn.

In considering the demographic* explosion—that is, the increase in population—in Europe in the years following the voyage of 1492, Crosby points to the spread of food crops out of the Americas.[2] He justifies his argument by observing that if you analyze the millions of calories produced per hectare of crop, American plants have a much higher yield than those grown in the Old World* (Africa, Europe, and Asia). Maize (corn), potatoes, and sweet potatoes (yams) all produced more than 7 million calories per hectare, while the only Old World crop to do this was rice, at 7.3 million calories. And manioc, a native plant of South America, produces 9.9 million. These foods became key parts of European and Asian diets, while cassava and maize fueled demographic growth in Africa (where they are now staples). Going in the other direction, Crosby points out that the livestock imported into the

Americas enabled the Western hemisphere to support many more people than it had previously. European populations that had grown rapidly—thanks to the spread of New World* crops—would later populate that same New World (that is, the Americas and the islands of the Caribbean) through immigration.

Although earlier works argue that political and technological factors, such as better weapons, drove the Spanish conquest of the New World, Crosby instead proposes that the devastating spread of the contagious viral disease smallpox* among previously unexposed populations played a more critical role in the submission of New World peoples to the conquistadors* (the Spanish conquerors). Crosby traces the effect of microbial disease on social structure and political organization—and theorizes that epidemics were as devastating indirectly (through economic and social collapse) as directly (through mortality). For example, he argues that the resistance shown by the indigenous Incan* people was weakened by a power struggle following the unexpected death of their ruler Huayna Capac,* almost certainly from disease. Crosby also connects this to his larger theme, which stresses the importance of understanding human history in its ecological context. In each chapter of *The Columbian Exchange* he examines a different aspect of history following Columbus's voyage, and shows how biological forces better explain historical phenomena than social and political factors.

Why Does *The Columbian Exchange* Matter?

In its arguments and execution, *The Columbian Exchange* ranks as one of the most important historical works of the past 50 years. Whereas historical writing overwhelmingly revolves around human events and movements, Crosby persuasively argues that the most significant consequences of Christopher Columbus's 1492 voyage were non-human.While this idea is now so widely accepted as to appear obvious, it was revolutionary when the book was first published in 1972.

Furthermore, *The Columbian Exchange* remains unparalleled as an example of how historical inquiry can be conducted by drawing on the sciences. The skill and creativity Crosby showed as he engaged with the scientific literature of the day reveals how much these fields can teach historians—and also proves that historians can contribute important ideas to epidemiology* (the study of the dynamics of epidemics of disease), biology, and other sciences. In consequence, Crosby's work deserves continued attention both as a significant historical text—foundational in the field of environmental history—and a stellar example of interdisciplinary scholarship (that is, as a work of scholarship that draws on the methods and aims of several fields of inquiry).

It follows that the central ideas of *The Columbian Exchange* have proved enormously influential—starting with the theory that epidemic disease drove the European conquest of the New World. Earlier scholars had explained the conquest as the result of European technology, or the American peoples' "primitive" religious and political systems. But in showing the effects of disease, Crosby undermines these older narratives and demonstrates the importance of biological factors. Crosby's "Columbian exchange" has become an important framework for organizing history into agreed periods. The impact of these ideas, together with their subsequent and lasting influence, have secured the place of Crosby's text among the most significant historical works.

The Columbian Exchange decisively reframed scholars' understanding of Christopher Columbus's 1492 voyage from Europe to the Americas and made a powerful case for the importance of environmental history in all fields of history. On the strength of these accomplishments, the work unlocked entirely new territories of historical inquiry.

---

1 Alfred W. Crosby, *The Columbian Exchange: Biological and Cultural Consequences of 1492* (Westport, CT: Praeger, 2003), xx.

2 Crosby, *The Columbian Exchange*, 165–207.

SECTION 1
INFLUENCES

THE AUTHOR AND THE HISTORICAL CONTEXT

KEY POINTS

* The American environmental historian Alfred W. Crosby's work demonstrates how non-human actors—plants, animals, and diseases, for example—shape history.

* Crosby became a critic of the "progress narratives" that characterized mainstream historical writing in the mid twentieth-century, according to which the study of history was a question of documenting human progress.

* Crosby wrote *The Columbian Exchange* as the environmentalism movement gathered steam in the 1960s and 1970s.

Why Read This Text?

Alfred W. Crosby's 1972 book *The Columbian Exchange: Biological and Cultural Consequences of 1492* argues that "the most important changes brought on by the Columbian voyages were biological in nature."[1] Crosby examines the implications of contact between the Eastern and Western hemispheres, starting with the voyage made in 1492 by the Italian explorer Christopher Columbus.* Earlier scholarship emphasized how cultural and technological factors shaped the relationships between hemispheres. Crosby contended, however, that non-human factors—among them the exchange of plants, animals, and microbes between the Old* and New Worlds*—mattered more.

Crosby's ideas were shaped by earlier works such as the

American historian Walter Prescott Webb's* *The Great Plains* (1931), which began to explore particular environments and ecosystems.[2] Crosby developed avenues opened in these works to produce a holistic* (that is, integrated and universal) view of how non-human, biological factors decisively shape and change human history—as shown in the aftermath of Columbus's voyage to the New World. While *The Columbian Exchange* is among several foundational texts in environmental history, it offers perhaps the most powerful example of environmental historians reframing popular notions of key historical moments by drawing both from science and disciplines such as geography and anthropology.

> "The 1970s was an encouraging period for environmer.talists, a decade that began in the United States with the first Earth Day and the formation of the Environmental Protection Agency and that spawned in the next few years the Clear Water Act, the Endangered Species Act and the Environmental Pesticide Control Act. Simultaneously, environmental history emerged as a separate and independent school of scholarship."
>
> —— Alfred W. Crosby, "The Past and Present of Environmental History"

Author's Life

Crosby received historical training as an undergraduate at Harvard University and a PhD in 1961 at Boston University under the direction of the American historian Robert E. Moody.* Despite his orthodox education, he quickly became a critic of mainstream historical scholarship. In the 2003 edition of *The Columbian Exchange*, Crosby described his teachers as men who rarely doubted the basic goodness of the society for which they had fought in World War II; they believed the world to be growing progressively better and that "the good guys always won."[3] But, for Crosby, the tumult of the 1960s raised doubts about these narratives and he began to consider approaches to history beyond the political, eventually conducting scholarship in the sciences and medical anthropology. Following a pair of articles in *The American Anthropologist* and the *Hispanic American History Review*, Crosby organized his writings on the consequences of Columbus's voyage into *The Columbian Exchange*. Crosby's teaching career took him to a number of schools—including Ohio State University, Yale, and the University of Helsinki—and he wrote and researched *The Columbian Exchange* during his time at Washington State University. As of 2015, Crosby is professor emeritus of history, geography and American studies at the University of Texas, Austin, where he has spent the bulk of his career.

Author's Background

The political and intellectual climate of the 1960s exerted a profound influence on Crosby's life and work. Environmentalism as a movement had been gaining momentum since the 1962 publication of the American marine biologist Rachel Carson's* book *Silent Spring*, which explored the ecological consequences of pesticide use. The Civil Rights movement* (in the course of which black Americans claimed their rights to equal treatment under the law), the Vietnam War,* and the general tumult of the decade shook his confidence in Western cultural superiority and the idea that history was a question of progress.[4] As a result, Crosby began to look outside the historical profession for solutions to his historical questions. He found his greatest influences in biology and a host of scientific disciplines: anthropology (the study of human cultures), geography, archeology, agronomy (the science of agricultural matters, including soil management), ecology, and other related fields. Acknowledging that the 1960s drove some to dogmatic political positions, he explained simply that "the sixties, which made ideologues of some, drove me to biology."[5]

He also began to study epidemiology* and to probe the importance of disease and food in human history. Once immersed in these subjects, Crosby stopped thinking in terms of the questions that had, until then, driven history as a discipline. As a result, he was free to think about radically different and often bigger issues. For example, Crosby wondered, "what kept people alive long enough to reproduce and what killed them?"[6] Once he began to wrestle with these questions, he saw the decisive influence of non-human factors in human history—and began a quest that would ultimately create an altogether-new lens through which to view historical events.

---

1 Alfred W. Crosby, *The Columbian Exchange: Biological and Cultural Consequences of 1492* (Westport, CT: Praeger, 2003).

2 Walter Prescott Webb, *The Great Plains* (Lincoln: University of Nebraska Press, 1981). For more

detail, see Alfred W. Crosby, "The Past and Present of Environmental History," *American Historical Review* 100, no. 4 (1995): 1177–89.

3 Crosby, *The Columbian Exchange*, xx.

4 Crosby, *The Columbian Exchange*, xx–xxi.

5 Crosby, *The Columbian Exchange*, xxi.

6 Crosby, *The Columbian Exchange*, xxi.

MODULE 2
ACADEMIC CONTEXT

KEY POINTS

* Alfred W. Crosby turned the attention of historians from politics toward the history of human relationships with the environment.

* Crosby used the tools employed by scientists to determine how human societies survived and perished in the face of environmental pressures.

* Crosby helped bring historians and scientists into a mutual conversation.

The Work in its Context

In 1972, when Alfred W. Crosby published *The Columbian Exchange: Biological and Cultural Consequences of 1492*, his approach to history was revolutionary. Until that point, traditional works focused on the political histories of specific nations. Humans had always stood at the center of historical scholarship; Crosby hoped to change that positioning. To make his case, he argued that a wider set of ecological relationships could decisively shape human history. He formed his theories by examining the non-human factors behind the radical changes that followed the Italian explorer Christopher Columbus's* 1492 voyage from Europe to the Americas.

In adopting this approach, Crosby hoped to solve vexing problems that had plagued earlier historians. Histories of the Spanish conquest* of the New World* focused on Spain's advantages in politics, culture, and technology but failed to account for the rapid decline of the Aztec* and Incan people of Central and South America. By placing epidemic disease at the center of the story of the Spanish conquest—the acute viral disease smallpox* in particular—Crosby fashioned a persuasive and original narrative to explain the scale and speed of the social and political collapse of indigenous American cultures.

Drawing on the aims and methods of several different fields of inquiry, the interdisciplinary nature of *The Columbian Exchange* formed the foundation of Crosby's success and its revolutionary approach. It was so unusual, however, that he had great difficulty in finding a publisher.[1] In the early 1970s, historical works that drew from the social sciences and humanities were far more common, which explains why an idea widely accepted today—that history and the hard sciences have important things to say to each other—was difficult to appreciate when Crosby first published *The Columbian Exchange*. The book went on to play a decisive role in changing this long-standing viewpoint.

> *"I fled from ideological interpretations of history and went in search of the basics, life and death ... what kept people alive long enough to reproduce, and what killed them? Perhaps food and disease?"*
>
> ——Alfred W. Crosby, preface to the 2003 edition of
> *The Columbian Exchange*

Overview of the Field

Crosby's approach had its precursors in the emerging field of environmental history, although his work diverged somewhat in ambition and scope. In 1893, the American historian Frederick Jackson Turner's* seminal paper "The Significance of the Frontier in American History" examined how the American frontier environment helped to promote individualism and democracy. Four decades later, the historian Walter Prescott Webb's* *The Great Plains* (1931) investigated change in particular environments and ecosystems.[2] Although Webb's work was particularly expansive, much of the scholarship that preceded *The Columbian Exchange* was narrowly bounded. Jackson and Webb's works embraced environmental themes more in their general subject matter than in applying sciences such as biology or ecology to spell out human relationships with the non-human world. Crosby certainly drew on these works, widening their approach into a holistic* view of the many ways plant, animal, and microbial factors exert a decisive influence on human history.

To achieve this goal, Crosby sought inspiration from a variety of fields ranging from geography to biology, becoming perhaps the first to bring scholarship in these areas—along with anthropology and epidemiology*—into the fertile ground of historical writing. In doing so, *The Columbian Exchange* raised questions other historians had neglected to ask, and proposed answers they had never considered. In his preface to the 2003 edition, Crosby compared posing the kinds of questions he asked to "replacing the standard film in your camera with infrared or ultraviolet film," concluding that it became possible to "see things you have never seen before."[3]

Academic Influences

With his writing principally influenced by the sciences rather than the historical profession itself, Crosby discovered altogether new vistas. In an article called "The Past and Present of Environmental History" (1995), Crosby cites work in several fields as essential in shaping his thinking and the field of environmental history as a whole. Archeologists were using new techniques to study ancient climates and ecosystems; ecologists such as Frederic E. Clements* and Charles Elton* broadened his views; and geographers such as Paul Vidal de la Blache* were developing theories that were "inspirational to historians with a burgeoning interest in the environment." For example,Vidal de la Blache's theory of possibilism,* with its argument that the environment does not completely decide the nature of human culture and society, replaced crude environmental determinism (according to which the environment does, indeed, determine everything) and pushed historians to examine human societal development within the finite scope of natural environments. Advances in epidemiology also influenced Crosby's work and for Crosby, the willingness of environmental historians to employ the hard sciences—a tendency he greatly shaped—makes the field distinct.

Crosby sums up the revolution: "Environmental historians have discovered that the physical and life sciences can provide quantities of information and theory useful, even vital, to historical investigation and that scientists try and often succeed in expressing themselves clearly."[4]

1 Alfred W. Crosby, "Reassessing 1492," *American Quarterly* 41, no. 4 (1989): 661.

2 Walter Prescott Webb, *The Great Plains* (Lincoln: University of Nebraska Press, 1981). For more detail, see Alfred W. Crosby, "The Past and Present of Environmental History," *American Historical Review* 100, no. 4 (1995): 1177–89.

3 Alfred W. Crosby, *The Columbian Exchange: Biological and Cultural Consequences of 1492* (Westport, CT: Praeger, 2003), xxi.

4 Crosby, "The Past and Present of Environmental History," 1189.

THE PROBLEM

KEY POINTS

- In *The Columbian Exchange*, Alfred W. Crosby sought to account for the "success" of European colonization in the New World.*
- Historians traditionally emphasized superior technology to explain how Europeans conquered Native American populations in the New World.
- Crosby stressed the importance of epidemic diseases as an agent of European colonization.

Core Question

Alfred W. Crosby's core aim, in *The Columbian Exchange: Biological and Cultural Consequences of 1492*, was to decenter the role of human factors in historical scholarship. Indeed, it was his consideration of non-human agents that allowed him to explain how European populations came to grow after Christopher Columbus's* voyage linked the New and the Old Worlds* in 1492. Previous scholarship had suggested urbanization, technological advances, and economic innovations as the key reasons. Crosby concludes that a much more important phenomenon was at work: the spread of higher calorie New World food plants to Europe.

Crosby also demonstrates that the post-1492 trend toward global biological homogeneity* (that is, a decrease in the variety of species) was more significant than the opening of new transatlantic trade routes. As a result of Columbus's voyage, previously separate populations of people, plants, and animals came into contact with each other—and decimated particular populations in the process. Diverse New World cultures were in part supplanted by imperial* colonial regimes, and the biological diversity of plants and animals gave way to large-scale monocultures (roughly, plantations of single crops) of cotton, sugar, tobacco, and so on. Crosby describes this trend as "one of the most important aspects of the history of life on this planet since the retreat of the continental glaciers."[1]

> "[The] trend towards biological homogeneity is one of the most important aspects of the history of life on this planet since the retreat of the continental glaciers."
> —— Alfred W. Crosby, *The Columbian Exchange*

The Participants

Earlier historians had argued that social and political factors drove history in the period following Christopher Columbus's 1492 voyage. As a notable example, the Belgian historian Charles Verlinden* traced the roots of European imperialism in the New World to Mediterranean history in the Age of the Crusades (the medieval period in which European nations organized military expeditions to the Middle East to "recover" Christian holy sites from Muslims) when "organizational structures and exploitative techniques that would be imposed on America in the sixteenth century were first tried."[2] The French historian Pierre Chaunu* likewise argued that the techniques of political and economic exploitation developed in the sugar islands of the Mediterranean paved the way for colonization in the Americas.

In sharp contrast, however, Crosby maintains in *The Columbian Exchange* that the most significant agents of change were biological. He outlines a "Columbian exchange" in which previously isolated plants, animals, and microbes came into contact with different peoples; the result was an overall trend towards global biological homogeneity. Whereas earlier works cite political and technological factors (including better weapons) as prime drivers in the Spanish conquest of the New World, Crosby proposes something altogether different. The devastating spread of smallpox* to previously unexposed populations, he argues, was more critical in explaining how New World peoples fell to the conquistadors.*

The Contemporary Debate

Crosby's *The Columbian Exchange* was not without its critics. In particular, scholars whose historical narratives stressed the importance of social and cultural factors took issue with the book. As Crosby remarked in his 1987 work *The Columbian Voyages, the Columbian Exchange, and Their Historians*, his approach to European colonization was "recondite [obscure] and discomforting" to those who embraced "Euro-American ethnocentrism"[3] (that is, those whose arguments automatically assumed the primacy of Euro-American perspectives).

One prominent opponent of Crosby's ecological approach was the American economic historian David Landes,* who argued that the "success" of European colonization stemmed from political, social, and cultural factors.[4] The opinions of critics like Landes embraced the importance of technological change—and observed that Columbus's voyage itself depended on social, political, and technological forces that enabled transatlantic travel in the first place. Responses of this kind constitute a key critical analysis of Crosby's work. Although mainstream academic thought has moved towards Crosby's perspective, the critiques of Landes and like-minded historians remain a powerful minority opinion. Environmental historians in subsequent decades, among them the American writers Richard White* and William Cronon,* have likewise argued that Crosby overstates the effect of biological factors on the course of history. While these historians accept the importance of non-human factors, they conclude that such factors are not entirely separate from human culture.

1 Alfred W. Crosby, *The Columbian Exchange: Biological and Cultural Consequences of 1492* (Westport, CT: Praeger, 2003), 3.

2 Crosby, *The Columbian Exchange*, 4–5.

3 Alfred W. Crosby, *The Columbian Voyages, the Columbian Exchange, and Their Historians* (Washington, DC: American Historical Association, 1987), 1–2.

4 David S. Landes, *The Wealth and Poverty of Nations: Why Some Are So Rich and Some So Poor* (New York: W.W. Norton, 1998).

THE AUTHOR'S CONTRIBUTION

KEY POINTS

* In *The Columbian Exchange*, Alfred W. Crosby sought to demonstrate the historical importance of non-human actors such as food crops, animals, and diseases.

* Crosby's focus on ecological factors provided a new way of understanding the success of European colonization* in the New World.*

* In using scholarship from the sciences to shed light on historical questions, Crosby helped to develop a radically new method of studying the past.

Author's Aims

In *The Columbian Exchange: Biological and Cultural Consequences of 1492*, Alfred W. Crosby sought to unite a wide range of scholarship in the humanities, social sciences, and biological sciences, with the goal of proving that "the most important changes brought on by the Columbian voyages were biological in nature."[1] Most if not all previous works emphasized human factors as causing large historical events such as the Spanish conquest of the Aztecs* (the inhabitants of what is today the country of Mexico) or the rapid economic growth of Europe after 1492. Instead, Crosby showed that the critical factors in these processes involved biological shifts—including the importation of new high-calorie foods such as potatoes, manioc, and corn to Europe—or epidemiological* effects, including the spread of smallpox* and other diseases.

Crosby also had a broader intent: to show that the study of non-human environmental factors is essential to the understanding of human history. He believed earlier historians had overplayed political and economic elements and so failed to grasp that humans can only be understood when viewed within their ecological context. To clarify this, he outlined how the accidental movement of pathogens* (organisms that cause disease) and other biota (the flora and fauna found in a specific place) could have bolstered the success of European empires in the Americas as much as deliberate human activity. Both materially and conceptually, Crosby hoped to prove that non-human forces act as the major drivers of human history.

> *"Virgin soil epidemics are those in which the populations at risk have had no previous contact with the diseases that strike them and are therefore immunologically almost defenseless. The importance of virgin soil epidemics in American history is strongly indicated by evidence that a number of dangerous maladies—smallpox, measles, malaria, yellow fever, and undoubtedly several more—were unknown in the pre-Columbian New World."*
>
> —— Alfred W. Crosby, "Virgin Soil Epidemics as a Factor in the Aboriginal Depopulation in America"

Approach

The intellectual framework of *The Columbian Exchange* rests on Crosby's fascination with works in the related fields of anthropology and geography, and in scholarship in the hard sciences: biology, geology, and medicine. Crosby cites the theories of ecologists such as the American scholar of plant evolution Frederic E. Clements* and the English zoologist Charles Elton* as particularly influential; they explored the impact of invasive species on natural ecosystems. His central concept—that the most significant effects of Christopher Columbus's* 1492 voyage to the Americas were biological—grew from his attempts to synthesize texts from outside history into a single work of historical scholarship.

Crosby built his overarching argument about the centrality of biological factors on several smaller conclusions. In fact, the theme of biology is common to the seemingly disconnected chapters of *The Columbian Exchange*, helping them find their coherence. Each explores one environmental consequence of contact between the hemispheres. This approach is powerful because, even though subsequent scientific research has cast doubt on some of Crosby's claims, he still achieves his overarching goal. Crosby's assertion that the Columbian exchange brought the sexually transmitted disease syphilis* to the Old World* is now disputed, as the balance of evidence indicates the disease probably already existed in Europe. Despite this, a larger, persuasive truth remains intact: when one unites evidence from anthropology, geography, biology, and other fields, non-human factors emerge as the decisive factor behind the changes brought about by the Columbian exchange.

Contribution in Context

Crosby's key themes in *The Columbian Exchange* apply research from the sciences in order to answer specific historical questions. But, to a degree, his aggressive stance on the importance of non-human factors within historical change was inspired by the environmentalist movement, first beginning to develop in the early 1970s. Although this approach would today be quickly identified as environmental history, it did not yet exist as a distinct subfield in 1972, the time that Crosby wrote *The Columbian Exchange*.

It would be incorrect to assume that Crosby's ideas in *The Columbian Exchange* were entirely original. Although he imports pre-existing ideas from the sciences into history, the result still remains highly innovative. The idea that research in the sciences could inform history was novel—especially the politically charged history that surrounded Columbus's 1492 voyage from Europe to the Americas. Furthermore, Crosby also pushes the reverse: that history can inform the sciences. This notion remains as boundary-shifting today as it was in 1972 because, while historians since have done a good job embracing the social sciences and humanities more broadly, engagement with the hard sciences has been much more limited. Crosby's willingness to draw from biology, medicine, and epidemiology*—the study of epidemics of disease—enabled and empowered him to make ambitious arguments about sweeping historical changes, even as he led sweeping changes in historical study.

---

1 Alfred W. Crosby, *The Columbian Exchange: Biological and Cultural Consequences of 1492* (Westport, CT: Praeger, 2003), xxvi.

SECTION 2
IDEAS

MAIN IDEAS

KEY POINTS

* The key themes of Alfred W. Crosby's *The Columbian Exchange* are the global spread of pathogens* (organisms that cause diseases) and food crops.

* Crosby's main argument was that the spread of European diseases to the New World* and the importation of New World crops into Europe paved the way for Europe's global hegemony in the centuries after Christopher Columbus's* voyage to the Americas in 1492.

* Crosby structured *The Columbian Exchange* around the spread of specific pathogens, plants, and animals.

Key Themes

Two key themes drive Alfred W. Crosby's *The Columbian Exchange: Biological and Cultural Consequences of 1492*: the global spread of food crops and of epidemic diseases. Within these themes, Crosby examines three distinct factors:

* the migration of diseases between the Old and the New World

* the impact of New World crops on the population of the Old World*

* the results of exchange between hemispheres of animals and non-food plants such as cotton and tobacco.

The Columbian Exchange takes a revolutionary look at the effect of epidemic disease on indigenous populations. Crosby argues that the devastating effects of disease, particularly smallpox,* help explain how Europeans came to dominate the New World. Drawing from his experience as a scholar of Latin America, Crosby recounts the conquistadors'* rapid conquest of Central and South America. As he explains, "We have for so long been hypnotized by the daring of the conquistador that we have overlooked the importance of their biological allies."[1]

Following his discussion of how microbes impacted the Columbian exchange, Crosby studies the role of plant and animal exchange between the Eastern and Western hemispheres. He maintains that the spread of food crops out of the Americas caused the Old World's population explosion in the years following 1492.[2] Combined, these two overarching themes allow Crosby to outline the specific ways that non-human factors decisively shaped global history after Columbus's voyage from Europe to the Americas that year.

> "The impact of the smallpox pandemic on the Aztec* and Incan empires is easy for the twentieth-century reader to underestimate. We have for so long been hypnotized by the daring of the conquistador that we have overlooked the importance of their biological allies."
> —— Alfred W. Crosby, *The Columbian Exchange*

Exploring the Ideas

Crosby traces the effect of microbial disease, making the compelling claim that these affected New World society and politics more than any invasion or military engagement. As Old World diseases moved into the New World, they not only had a direct effect on mortality, indirectly they also caused devastation as they ignited the social and political collapse of native populations. Crosby contends that the Incas,* for example, did not so much yield to the military might of the Spanish as succumb to the diseases they brought with them. The unexpected death of their ruler, Huayna Capac,* almost certainly stemmed from disease and led to chaos and weakness in their attempts to resist the invading armies. Crosby also connects this to his larger theme that human history is shaped by the forces of global ecology.

Crosby reasserts this theme in a subsequent chapter on the sexually transmitted disease syphilis,* which he contends moved from the New World to Europe with devastating effects. While not as disastrous as the smallpox epidemic in the Americas, syphilis had a widespread impact in Europe, and Crosby teases out many cultural ripples in his discussion. Although recent research into the transmission of syphilis suggests Crosby was incorrect in claiming that the disease originated in the New World, his now-disputed argument does not undermine his central theme concerning the spread of diseases through the Columbian exchange.

Crosby also argues that the spread of food crops out of the Americas caused the Old World's post-1492 population explosion.[3] To back up his argument, Crosby bypasses the forces many historians cite—whether in the spheres of politics, conflict, or colonization—and looks instead at the food that crossed from the Americas back to Europe. American plants such as maize (corn), potatoes, and sweet potatoes (yams) all have much denser calorie counts than the crops eaten by Europeans and, in time, these foods became key parts of European and Asian diets. In the opposite direction, livestock imported into the Americas enabled the Western hemisphere to support a burgeoning population of European immigrants. European populations thrived thanks to the spread of New World crops, and became a force of change as they moved into the Americas. Crosby points out, however, that this movement of Old World people would never have taken place if not for the earlier movement of New World crops to their shores.

Language and Expression

In asserting the historical importance of ecological exchanges, Crosby goes beyond food-crop transfers to build his larger argument around the themes of disease and epidemics. He persuades us that non-human factors played a central role in the Columbian exchange by highlighting example after example where plants, animals, or disease drove major changes once attributed to technological change, political superiority, or cultural forces. This structure, which supports his argument well, arises from the stitching together and expansion of two previously published articles.

Crosby's *The Columbian Exchange* addresses traditional historiographical* questions—that is, roughly, questions pertinent to the field of historical research—by using innovative methods. Although Crosby's book would today quickly be identified as a work of environmental history, when he first published *The Columbian Exchange*, this was a field with no distinct identity— arguably, it did not even exist. Thanks to the way he engaged and reframed debates previously explained in political or economic terms, it was a discipline he would help to spawn. In this way, *The Columbian Exchange* provided new answers to familiar questions such as "How did the Europeans conquer the New World?" and "Which factors shaped the rapid economic and demographic* growth of the Old World after 1492?"

The Columbian Exchange also challenged historians to recognize the relevance of knowledge straight from the sciences, forcing them to wade into unfamiliar waters where they would encounter and discover the intellectual currents of archeology, biology, and epidemiology.

---

1 Alfred W. Crosby, *The Columbian Exchange: Biological and Cultural Consequences of 1492* (Westport, CT: Praeger, 2003), 52.

2 Crosby, *The Columbian Exchange*, 52.

3 Crosby, *The Columbian Exchange*, 165–207.

MODULE 6
SECONDARY IDEAS

KEY POINTS

* Alfred W. Crosby examined the intellectual impact of Christopher Columbus's* 1492 voyage to the New World* and presented a pessimistic assessment of the environmental legacy resulting from European conquest.

* In exploring the intellectual impact of the Columbian exchange, Crosby provided subsequent historians with a model to examine the interplay between ideas and the material world.

* Since the publication of *The Columbian Exchange*, historians have acquired a greater understanding of the negative consequences of Columbus's voyage to the Americas.

Other Ideas

One important subordinate idea developed by Alfred W. Crosby in *The Columbian Exchange: Biological and Cultural Consequences of 1492* involves the conceptual effect the Columbian exchange had on intellectual systems on both sides of the Atlantic. In the book's first chapter, "The Contrasts," Crosby explains how the discovery of new peoples, crops, and animals had to be squared with the fundamental beliefs that Europeans and indigenous people held about the world in which they lived. In Spain, for instance, sixteenth-century religious thinkers such as José de Acosta* suggested that multiple creations explained the origins of those native to South America. This is especially interesting because it hints at the relationship between the natural world and the realm of ideas, but it is suggested rather than fully developed.

While it is debatable how much Crosby's pessimism pervades his book, it nonetheless represents another important subordinate theme of *The Columbian Exchange*. That it dominates the final essay is quite clear; Crosby closes by telling us that "the Columbian exchange has left us with not a richer but a more impoverished genetic pool.We, all of the life on this planet, are the less for Columbus, and the impoverishment will increase."[1] Crosby's point is that the circulation of peoples, plants, animals, and microbes between the hemispheres created transformations that tended toward biological homogeneity,* squashing human and biological diversity while spreading sameness.

> "Columbus was the advanced scout of catastrophe for Amerindians. There were a few happy sequelae—the flowering of equestrian cultures in the American grasslands, for instance—but on balance, the coming of whites and blacks brought disease, followed by intimidation, eviction, alcoholism ... and obliteration of many peoples and ways of life."
>
> ——Alfred W. Crosby, *The Columbian Voyages, the Columbian Exchange, and Their Historians*

Exploring the Ideas

Crosby's pessimistic assessment is not limited to the book's conclusion; he develops this important sub-theme throughout the text. His claim that "we, all of the life on the planet, are the less for Columbus" was meant to draw attention to the previously neglected ecological and human toll of the explorer's voyages. Crosby contended that increased biological homogeneity and decreased global biodiversity* (the diversity of all the world's many species) represented a key global shift following Columbus's 1492 voyage from Europe to the Americas. Human cultures were threatened or completely eliminated by diseases, just as particular plant and animal populations shrank, disappeared, or expanded—all depending on their role in a newly global economy.

The book's pessimism, which has proved controversial, is counter to perceptions of Columbus as one of history's great heroes. Yet it is fundamental to Crosby's broad point about the importance of non-human factors to the Columbian exchange. The eradication of Amerindian human populations paved the way for the global dominance of European civilization; as Amerindians succumbed to diseases to which they had no natural immunities, it reduced their ability to resist European colonists. As Crosby remarked, "When strangers meet, the degree of difference between their bacterial florae can make more history than the differences between their customs."[2]

Overlooked

While most of *The Columbian Exchange* has been carefully studied, there are nonetheless underappreciated aspects of the text. Perhaps the most neglected part of *The Columbian Exchange* is chapter one, "The Contrasts," in which Crosby weaves a discussion of the differences between the Eastern and Western hemispheres. He tackles this through an examination of "the problem of America"—that is, the effect of Christopher Columbus's "discovery" of the Americas on the European imagination.[3] Though fascinating, this aspect of the text is underdeveloped. As opposed to a thorough analysis, the discussion feels like a series of interesting but disconnected observations about the European imagination. As Crosby wrote, "America was a very square peg to fit into the round hole of Genesis."[4]

Nevertheless, Crosby's investigation of how the New World changed Europeans' ideas about themselves and the environment proved ahead of its time. Though Crosby himself suggests that environmental historians focus too much on material reality and too little on ideas about the environment, his brief discussion in chapter one does consider the world of ideas. Crosby ponders how the "discovery" of the Americas raised questions about Christianity's creation myth, spurring some unorthodox thinkers to conclude that there must have been multiple creations. Rather than merely address the environment itself, Crosby's opening chapter provides insight into ideas of the environment—and how to think about them. This compelling viewpoint reminds us just how much the New World's species and its people impacted on European intellectual and spiritual frameworks, including Christian thought.

1 Alfred W. Crosby, *The Columbian Exchange: Biological and Cultural Consequences of 1492* (Westport, CT: Praeger, 2003), 219.

2 Alfred W. Crosby, *The Columbian Voyages, the Columbian Exchange, and Their Historians* (Washington, DC: American Historical Association, 1987), 24.

3 Crosby, *The Columbian Exchange*, 12.

4 Crosby, *The Columbian Exchange*, 12.

ACHIEVEMENT

KEY POINTS

* "Columbian exchange" has become a phrase in common use by historians and scholars in the social sciences.

* *The Columbian Exchange* was published as environmental historians sought to analyze the relationships between humans and the natural environment.

* Since the 1970s, economists have attempted to measure the impact of the Columbian exchange in quantitative terms—that is, through measurements, including statistics.

Assessing the Argument

Though Alfred W. Crosby's 1972 book *The Columbian Exchange: Biological and Cultural Consequences of 1492* remains as relevant as ever, it has started to appear dated for a curious reason: Crosby's argument has been largely accepted by academics. What was once a revolutionary approach is now common wisdom and environmental historians eagerly engage with scientific research and ecology as they explore humanity's relationship with the environment. While historians and scholars today debate the exact role of epidemic disease in the European conquest, they generally accept Crosby's claim that it holds central importance. Similarly, scholars generally agree with Crosby's claims about the importance of New World* foods to Old World* populations, and the debate has moved on to economists' attempts to quantify them more precisely.[1] Thus the current consensus view holds that Crosby's analysis in *The*

Columbian Exchange, while broadly correct, can be pushed further.

Meanwhile, *The Columbian Exchange* continues to be considered a pioneering work in the field of environmental history. Crosby's approach was so successful that the idea of the "Columbian exchange" now offers a generally accepted framework for making sense of the changes after Christopher Columbus's* 1492 voyage from Europe to the Americas: in fact, scholars freely use the phrase without directly citing Crosby. While many of his specific arguments have lost some relevance because they are so well accepted, his approach remains important and relevant.To quote from the environmental historian J. R. McNeill's* foreword to the 2003 edition of *The Columbian Exchange*, Crosby's legacy "lies not in the comprehensiveness of chronicling the Columbian Exchange, but in the establishment of a perspective, a model for understanding ecological and social events."[2]

> *"Crosby's legacy lies not in the comprehensiveness of chronicling the Columbian Exchange, but in the establishment of a perspective, a model for understanding ecological and social events."*
>
> ——J. R. McNeill, foreword to the 2003 edition of
> *The Columbian Exchange*

Achievement in Context

It is impossible to overstate the novelty of Crosby's argument and methodology when *The Columbian Exchange* was published in 1972. At the time, Crosby's ideas, themes, and methods positioned the book outside mainstream history. Crosby had great difficulty finding a publisher and, when *The Columbian Exchange* finally appeared in print, it could have more appropriately qualified as a work of anthropology or medical history. Yet the book itself, and its eventual percolation throughout the historical realm, has meant that Crosby's ideas now lie at the heart of the field of history. This position would likely have shocked scholars active when the work was first published.

The Columbian Exchange challenged historical writing bound by the cultural and political assumptions prevalent prior to the publication of the text. In criticizing those assumptions, Crosby proposed that universal biological questions held prime importance in driving human history, an approach which served to decenter Eurocentric* narratives that tied the global supremacy of European civilization to its superiority in politics, culture, intellect, and technology. Instead, Crosby revealed that Europe was able to colonize the Americas—and the rest of the world in later centuries—more as a result of an accident of history and a convergence of biological factors.

Limitations

Though the influence of *The Columbian Exchange* outside the historical profession has been limited, the book has produced two important results. First, as we have seen, the Columbian exchange is now the generally accepted framework for thinking about Christopher Columbus's 1492 voyage; and, second,

Crosby's emphasis on non-human factors in history has inspired similar approaches in the humanities and social sciences. Scholars in economics, geography, and related fields use the phrase "Columbian exchange" without direct acknowledgement of Crosby's work, a clear sign of the idea's pervasiveness.

Though Crosby's work drew directly from the sciences, and benefited from their impact, it is more difficult to trace the reverse: how it has influenced fields such as biology and epidemiology.* In a sense, those fields were already asking the kinds of questions Crosby hoped to place at the center of historical inquiry. Yet many of the sciences, and especially fields such as public health, have embraced a more historical approach. Thus it is plausible to suggest that Crosby's book might have had an influence.

In the social sciences beyond history, *The Columbian Exchange* has produced a clearer impact. Recent work in economics has tried to quantify the importance and effect of the Columbian exchange on demographic* growth. In an article from the *Journal of Economic Perspectives,* the economics professors Nathan Nunn* and Nancy Qian* note that the study of biological effects from the Columbian exchange has been "neglected" in economics literature, and they provide their own analysis.[3] In their findings, which largely corroborate Crosby's work, they notice important links between New and Old World prices for products that eventually became staples, such as sugar and coffee. Similar work has attempted to quantify the extent of epidemic disease in the New World. Much of this work has tried to add quantitative evidence to Crosby's abstract claims. Indeed, some assessments of Crosby's thinking could change—as happened in the case of his claims about the origin and spread of syphilis.* But, to date, Crosby's arguments have been largely corroborated, and hold sway beyond the field of history.

---

1 Nathan Nunn and Nancy Qian, "The Columbian Exchange: A History of Disease, Food, and Ideas," *Journal of Economic Perspectives* 24, no. 2 (2010): 163–88.

2 J. R. McNeill, foreword to *The Columbian Exchange: Biological and Cultural Consequences of 1492*, by Alfred W. Crosby (Westport, CT: Praeger, 2003), xiii.

3 Nunn and Nancy, "The Columbian Exchange," 164.

MODULE 8
PLACE IN THE AUTHOR'S WORK

KEY POINTS

* The overall thrust of Crosby's work has been to examine how non-human actors (notably plants, animals, and diseases) shaped the course of historical events.

* Alfred W. Crosby's *The Columbian Exchange* set the agenda for his subsequent work.

* *The Columbian Exchange* established Crosby as one of the founding figures of the discipline of environmental history.

Positioning

Alfred W. Crosby's doctoral dissertation *America, Russia, Hemp, and Napoleon* (published as a book in 1965) broached radically different themes to *The Columbian Exchange: Biological and Cultural Consequences of 1492*, examining the politics of trade—particularly in hemp—between Russia and the United States in the years between the American Revolution* and the War of 1812* (both armed conflicts in which the young United States fought Great Britain for its political and economic independence). Although his dissertation was an example of the very kind of political history he would dismiss in future years, glimpses of the later Crosby appear: his subject is large and he traces unexpected connections between disparate places and political events. It would take seven years—and a new decade—before Crosby completed the jump to ecological history.

    The Columbian Exchange grew out of two previously

published articles, portions of which Crosby used in the book itself. The second chapter, on epidemic disease and Spanish conquest, ran in a journal called the *Hispanic American Historical Review.* Parts of the fourth chapter, on syphilis,* first appeared in *American Anthropologist.* Both articles are largely the same as the book chapters, which helps explain the book's somewhat disconnected structure: six relatively independent essays unite to form Crosby's larger argument. Starting with his exploration of diseases in the aftermath of 1492, Crosby fleshed out *The Columbian Exchange* by attempting to trace all of the effects of Columbus's* voyage— while giving readers a better understanding of the biological causes behind them.

> *"Twenty years ago I finished a book on the impact of the Columbian voyages on the peoples of the world ... I had a hard time finding a publisher, and almost gave up on the book before Greenwood Press spontaneously wrote me to ask if I had anything publishable on hand ... the book sold and continues to sell modestly but steadily, three thousand or so a year for a total that must be upwards of forty or fifty thousand by now."*
>
> —— Alfred W. Crosby, "Reassessing 1492"

Integration

As Crosby's first significant book publication, *The Columbian Exchange* set the agenda for questions he would explore throughout his career. His subsequent works have examined how humans exist within a broader web of ecological relationships.

Following *The Columbian Exchange*, Crosby continued to ask questions that spanned centuries of global change. *Ecological Imperialism: The Biological Expansion of Europe, 900–1900* makes an argument that covers 1,000 years of human history.[1] In this seminal 1986 work, Crosby pursues the argument first developed in *The Columbian Exchange* that it was biology rather than military might that spurred the success of European empires— but dramatically expands the chronological scope. Crosby stretches back to the invasions of Norseman and the Crusades, and examines the making of "Neo-Europes" in the Pacific during the nineteenth century. He also offers a critical examination of the Columbian voyages' historiography* in his short 1987 book *The Columbian Voyages, the Columbian Exchange, and Their Historians*.

Thematically, Crosby's corpus is unified; it analyzes on a large scale the exchange of people, plants, animals, foods, and diseases between hemispheres. He has, however, expanded his geographic scope over the years. In *The Columbian Exchange*, Crosby primarily studies the Americas. In *Ecological Imperialism*, he places the expansion of European disease, animals, and agriculture at the center of global European imperialism.* And his interest in disease, clearly evident in *The Columbian Exchange*, continues in subsequent works such as *America's Forgotten Pandemic*,[2] his influential history of the influenza pandemic of 1918* (a mass outbreak of influenza that infected nearly 500 million people worldwide and killed between 50 and 100 million).[3]

Significance

The Columbian Exchange has been so influential that, in the light of the historical literature that has followed, it is difficult to gauge quite how distinctive and original Crosby's arguments were. Many of Crosby's assertions about the importance of non-human factors in history have turned into fundamental assumptions for mainstream historians.

Although the work stands as one of several foundational texts in environmental history, it also provides perhaps the most powerful example of how environmental historians can reinterpret key historical moments by drawing both from the sciences and from disciplines such as geography and anthropology.

The central ideas of *The Columbian Exchange* have proved influential both in their specific claims and on the greater whole of historical scholarship. The idea, for example, that epidemic disease was the decisive factor in the European conquest of the New World* has been enormously significant. Earlier scholars had put greater weight on European technology and the "primitive" nature of indigenous American religious and political systems as root causes. But, in showing the effects of disease, Crosby undermined these older narratives and bolstered the argument for the importance of non-human factors. Similarly, Crosby's emphasis on a "Columbian exchange" framework resulted in a dominant historical paradigm—that is, a conceptual model—that explained a significant period of history like no previous publication.

The impact created by these ideas on their first appearance, and their subsequent influence in succeeding years, have secured the place of Crosby's text in the list of great historical works.

---

1 Alfred W. Crosby, *Ecological Imperialism: The Biological Expansion of Europe, 900–1900*, 2nd edn (Cambridge: Cambridge University Press, 2004).

2 First published in 1976 as *Epidemic and Peace* by Greenwood Press, Westport, CT.

3 Alfred W. Crosby, *America's Forgotten Pandemic: The Influenza of 1918*, 2nd edn (Cambridge: Cambridge University Press, 2003).

SECTION 3
IMPACT

THE FIRST RESPONSES

KEY POINTS

* Early critics of *The Columbian Exchange* rejected Alfred W. Crosby's claim that biological factors were primarily responsible for the dominance of Western civilization.

* In response to empirical criticism of his work (that is, criticism based on deduction made from observable evidence), Crosby has conceded that syphilis* was not introduced into Europe from the Americas.

* Crosby's work has been supported by a new generation of historians seeking alternative explanations for the global hegemony of European civilizations.

Criticism

The bulk of early criticism directed at Alfred W. Crosby's *The Columbian Exchange: Biological and Cultural Consequences of 1492* addressed the deep pessimism of the book's final essay. "We, all of the life on this planet," he observes, "are the less for Columbus."*[1]

Crosby contended that a key global shift following Christopher Columbus's 1492 voyage from Europe to the Americas was the simultaneous rise in biological homogeneity* and a decline in global biodiversity.* Many reviewers latched onto this claim as overstated and problematic. The geographer Gary S. Dunbar,* professor emeritus at the University of California at Los Angeles, observed that the book "concludes on an unduly pessimistic note."[2]

Here one finds the first hints of how Crosby's book stirred up controversy: his assertion that the traditional narratives of European global dominance had been overly celebratory, sidestepping the brutality of European supremacy. Columbus had (and still has) a national holiday named after him in the United States—and Crosby was telling whoever would listen that Columbus's voyage had made everyone in the world worse off.

This early criticism of Crosby's pessimism foreshadowed the future direction these debates would take. In response to thinkers such as Crosby and the best-selling geographer Jared Diamond,* who tried to explain European dominance as being chiefly the result of non-human factors, many scholars celebrated culture as central to economic and political success. The American economic historian David Landes* challenged Crosby's viewpoint by backing political, social, and cultural factors as key determinants of success or failure.[3] Yet, while the early reception of *The Columbian Exchange* hinted at the debate to come, it would be inaccurate to characterize it as immediately heated. In the early 1970s, scholars had yet to appreciate the full strength of Crosby's argument.

> "Whereas thirty years ago Crosby's ideas met with indifference from most historians, neglect from many publishers, and hostility from at least some reviewers, they now figure prominently in conventional presentations of modern history."
>
> ——J. R. McNeill, foreword to the 2003 edition of *The Columbian Exchange*

Responses

Crosby's response to criticism of *The Columbian Exchange* is apparent in both the opening of his subsequent book, *Ecological Imperialism* (1986), and his preface to the 2003 edition of *The Columbian Exchange*. Crosby admitted that, in light of new scientific research, he was mistaken about the origin of syphilis and revised his claims as a result. In a sense, this criticism reflected what was special about his book; because Crosby engaged so thoroughly with scientific research, particular claims merited notice and could be disproved. Crosby also acknowledged, in his 2003 preface, that his account of the ways in which disease epidemics paved the way for European colonization* lacked enough nuance to include non-biological factors. In fact, he had long since produced a more subtle account in a 1976 article "Virgin Soil Epidemics as a Factor in the Aboriginal Depopulation in America."[4]

In other areas, however, Crosby stuck to his guns. He continued to reject the idea that European superiority had more to do with cultural and political organization than non-human factors. In fact, his book *Ecological Imperialism* further attacks this idea. Crosby traces the development of what he calls "neo-Europes" in the New World,* asserting that biological and ecological processes were at the heart of imperialism.* While competing explanations of European dominance in a world marked by the colonialism that followed Columbus's voyage of 1492 remain unresolved, Crosby's approach is the widely accepted account within the historical profession.

Conflict and Consensus

A number of Crosby's empirical conclusions have been revised and rejected, starting with his deduction that the Columbian exchange brought syphilis to Europe. The balance of evidence now indicates that the disease already existed in Europe prior to 1492. Yet the larger point of his work remains persuasive: in combining evidence from anthropology, geography, biology, and other fields, non-human factors emerge as the decisive agents of change brought about by the Columbian exchange.

More recent work accepts the importance of non-human factors (which, for Crosby, means plants, animals, and diseases) while arguing that they cannot be treated as wholly separate from human culture.[5] This literature prefers to blur the boundaries between nature and culture by suggesting how they fundamentally intertwine—a position Crosby hints at but does not directly emphasize in his chapter on epidemic disease and New World conquest. In his article "Virgin Soil Epidemics as a Factor in the Aboriginal Depopulation in America," Crosby explicitly argues that the effects of epidemics are not wholly biological, but must instead be understood as arising from a mix of human and non-human factors.[6]

Within the context of his entire corpus, Crosby's later works serve more to advance his early ideas than directly respond to critics of *The Columbian Exchange*. In fact, Crosby has responded to assessments of *The Columbian Exchange* by incorporating constructive criticisms into his work, while continuing to attack claims for European dominance that focus on the nature of European culture.

1 Alfred W. Crosby, *The Columbian Exchange: Biological and Cultural Consequences of 1492* (Westport, CT: Praeger, 2003), 219.

2 G. S. Dunbar, "Review of *The Columbian Exchange,*" *William and Mary Quarterly* 30, no. 3 (1973): 543.

3 David S. Landes, *The Wealth and Poverty of Nations: Why Some Are So Rich and Some So Poor* (New York: W.W. Norton, 1998).

4 Alfred W. Crosby, "Virgin Soil Epidemics as a Factor in the Aboriginal Depopulation in America," *William and Mary Quarterly* 33, no. 2 (1976): 289–99.

5 See William Cronon, *Changes in the Land: Indians, Colonists and the Ecology of New England* (New York: Hill and Wang, 1983); and Richard White, *The Roots of Dependency: Subsistence, Environment and Social Change Among the Choctaws, Pawnees and Navajos* (Lincoln: University of Nebraska Press, 1983).

6 Crosby, "Virgin Soil Epidemics," 289–99.

MODULE 10
THE EVOLVING DEBATE

KEY POINTS

- Alfred W. Crosby's ecological approach has significantly impacted the historiography* of European colonialism.
- "The Columbian exchange" has become a central concept in environmental history and the history of European imperialism.*
- *The Columbian Exchange* is one of the foundational works of environmental history.

Uses and Problems

Scholars following Alfred W. Crosby have extended his work in important ways. *The Columbian Exchange: Biological and Cultural Consequences of 1492* was a needed corrective since, at the time of publication, historical writing revolved mostly around political or social factors. More recent work has tried to address the interplay between human and non-human aspects, an approach Crosby only hints at in *The Columbian Exchange* but embraces more explicitly in his later writing. In fact, Crosby acknowledges that his emphasis on non-human factors was a starting point for today's scholarship, which tries to break down the boundary between the human and non-human. For example, this literature has agreed that the disease of smallpox* played a decisive role in European conquest—noting, however, that the effects of the smallpox microbe working within a social and political structure made the epidemic deadly. As a result, these works more forcefully claim that all epidemics are produced by biological and cultural factors working in concert.

Along these lines, scholars have first and foremost pursued a more nuanced understanding of how biology and humanity feed off each other in a historical framework. Crosby intended *The Columbian Exchange* to serve as a remedy of sorts—he was trying to persuade historians to leave behind their cherished notions of politics and society, and consider the importance of disease and of new types of crops. Although to that end he succeeded, later historians have had to develop stronger theoretical frames of reference. These newer approaches break down divides between nature and culture as arbitrary, and question Crosby's supposition that human and non-human factors can be cleanly separated. However, more theoretically subtle and unified accounts, such as the American environmental historian Richard White's* *The Organic Machine* (1995), would likely not have been possible without Crosby's foundational work.

> *"The Columbian Exchange* has provided economists interested in the long-term effects of history on economics with a rich historical laboratory. Economic studies have thus far mainly focused on how European institutions, through colonialism, were transplanted to non-European parts of the world."
> —— Nathan Nunn and Nancy Qian, "The Columbian Exchange: A History of Disease, Food, and Ideas"

Schools of Thought

The Columbian Exchange is better known for updating the methods used by historians than inspiring "followers." The historians who established environmental history as a mainstream field during the late 1970s and 1980s—among them, Patricia Limerick,* J. R. McNeill,* Richard White, and Donald Worster*—could not really be described as Crosby's disciples. These historians engaged with themes Crosby addressed and drew from other, older, scholars such as the American environmental historian Walter Prescott Webb.* In that sense, *The Columbian Exchange* proved a key text in a broader trend towards environmental history.

Crosby's perspective is ecological, and this particular approach to history has a number of followers, notably White and the influential environmental historian William Cronon.* Both White's *The Organic Machine*[1] and Cronon's *Changes in the Land* (1983)[2] focus, like *The Columbian Exchange*, on a historical consideration of the ecological relationships—plant, animal, and microbial—enmeshed with human beings and societies. Although the American environmentalist Aldo Leopold* first formulated the idea of writing history ecologically, Crosby was the first academic historian to take up the challenge. Just as Crosby influenced ecological history, he also played a decisive role in encouraging historians to engage with scientific literature.

In Current Scholarship

Crosby's *The Columbian Exchange* decisively reframed the ways in which scholars understood the consequences of Christopher Columbus's* voyage in 1492. His book made a powerful case for the importance of environmental history to all fields of historical inquiry—and, in the process, exerted profound influence on environmental historians. Today, these scholars engage disciplines as diverse as geology and epidemiology,* and the Columbian exchange is now a recognized historical phenomenon. Historians no longer question whether an exchange took place, but try to assess and pinpoint its reach. More recent historical work has sought to better theorize how biological factors and social practices interact, while economists have quantified the extent of the exchange.[3]

Although they may disagree with some of the specifics, today's environmental historians broadly accept the central premises of *The Columbian Exchange.* It would, however, be inaccurate to say that it created a unified school or approach. The work did not seek to create a following of "Crosbians;" rather, it exemplified a specific way of examining history. And this way, with its emphasis on ecological and environmental matters, was already crystallizing when *The Columbian Exchange* was published.

If few of the scholars inspired by the work would consider themselves Crosby disciples, environmental history as a field owes a great deal to him nonetheless—just as it owes a significant debt to earlier scholars and thinkers such as Walter Prescott Webb, Frederick Jackson Turner,* and Aldo Leopold.

---

1 Richard White, *The Organic Machine: The Remaking of the Columbia River* (New York: Hill and Wang, 1995).

2 William Cronon, *Changes in the Land: Indians, Colonists and the Ecology of New England* (New York: Hill and Wang, 1983).

3 Nathan Nunn and Nancy Qian, "The Columbian Exchange: A History of Disease, Food, and Ideas," *Journal of Economic Perspectives* 24, no. 2 (2010): 163–88.

MODULE 11
IMPACT AND INFLUENCE TODAY

KEY POINTS

* Alfred W. Crosby's *The Columbian Exchange* is considered a pioneering work in the field of environmental history.

* Crosby's book still poses a challenge to narratives that explain the dominance of Europe in the years following 1492 by pointing to cultural, political, and social factors.

* Opponents of Crosby's approach stress that cultural factors are more critical in explaining European success and power in the early-modern period (roughly, from the end of the fifteenth to the end of the eighteenth century).

Position

The specialized questions raised in Alfred W. Crosby's 1972 work *The Columbian Exchange: Biological and Cultural Consequences of 1492* remain a subject of debate, primarily within the academic world. Although scholars have broadly accepted Crosby's claims, questions about the specific impact of New World* foods or the importance of Old World* disease linger—inspiring both deep inquiry and pointed criticism. While the concept of a Columbian exchange is now an accepted framework for interpreting significant historical features of the last 500 years or so, Crosby's larger point about the importance of non-human factors in explaining European power and its conquest of the Western hemisphere has some notable academic critics.

Although Crosby's argument, once considered revolutionary,

broke down paradigms (that is, interpretive models) when it was introduced in the early 1970s, more recent work has challenged his thesis on theoretical grounds. It has been observed that Crosby treats culture and nature as though historians can consider them separately when, in reality, they are interrelated. Furthermore, it has been contended, Crosby's discussion of the importance of the Columbian exchange on European and American intellectual history is underdeveloped. Recent work has more forcefully emphasized the relationship between ideas and environment. Within the context of environmental history in the twenty-first century, *The Columbian Exchange* is a victim of its own success. It has become so thoroughly incorporated into the study of history that its once-revolutionary ideas seem either clichéd or lacking the nuance of more recent work.

> *"At the time of publication, Crosby's approach to history, through biology, was novel ... Today,* The Columbian Exchange *is considered a founding text in the field of environmental history."*
>
> —— Megan Gambino, "Alfred W. Crosby on the Columbian Exchange"

Interaction

Many of the key arguments in *The Columbian Exchange* are now widely accepted and, within the historical profession, Crosby remains little challenged, in part because he has had so much influence on environmental historians. But certain aspects of his thought remain unpopular in various corners of academia. Specifically, these objections concern the implications of Crosby's argument, which implicitly attacks cultural explanations for European global dominance. Suspicion of Crosby's stance has come from conservative academics, and particularly from scholars in fields such as economics. They prioritize the development of European government and financial institutions to explain change in the early-modern period.

The Columbian Exchange has also been criticized as too material, in that it marginalizes the importance of human perceptions and attitudes about the environment. Crosby himself has observed that scholars like him tend "to be more interested in dirt than in perceptions, per se, of dirt. They have no doubts about the reality of what they deal with, nor about their ability to come to grips with it."[1] More recent scholarship contends that this is a significant oversight, since human perceptions of the environment and ecology have served as important forces in human history.

The Continuing Debate

Crosby's move to decenter historical narratives that view the success of European imperialism* as a result of superior technology, society, and politics remains disputed. Though many scholars agree with Crosby, a number of others disagree, particularly in the field of economic history. This disagreement is most obvious when we contrast two recent works by the American geographer Jared Diamond* and the American economic historian David Landes.* In his popular book *Guns, Germs and Steel* (1997), Diamond—

an indirect disciple of Crosby—advances an argument quite similar to that proposed in *The Columbian Exchange*. Geography and the natural world, Diamond contends, go a long way towards explaining the success or failure of particular peoples and countries.

Landes, however, in his *The Wealth and Poverty of Nations* (1998), argues that cultural factors are more relevant for explaining European success and power in the early-modern period. Although the weight of scholarship seems to support Crosby and Diamond's side of the debate, in the field of economics a great deal of quantitative scholarship (that is, scholarship drawing on measurable evidence such as statistics) supports Landes's account. Though unpopular with historians, *The Wealth and Poverty of Nations* also enjoys popular support from people who prefer to understand European success as the result of Western cultural values.

While Crosby's academic critics have reasoned arguments in their favor, much of the public conversation is heavily politicized. Crosby himself explains that his work grew out of the turmoil of the 1960s and disillusionment with his conservative predecessors. Generally, defenders of cultural arguments for European dominance are politically conservative, believing in the innate superiority of European civilizations. As the historian Margaret Jacob* has remarked, the very question of the West's success "has become unfashionable on the left ... it has been consigned, broadly speaking, to the right."[2] As a result, public conversations on these issues reflect sharp polarization. People on the political left view cultural arguments as Eurocentric* (founded on an unspoken assumption of European superiority) and as justifications for the brutality of colonialism. Meanwhile, right-leaning critics argue that crediting European dominance to nature or chance denies the values that Western democracies hold dear.

The public discourse lacks some level of sophistication, and sharp disagreement divides the two sides on the issue of nature versus culture. But both sides of the academic debate would likely argue that the reality reflects, to some extent, a combination of the two—if they were ever motivated, that is, to hear each other out.

---

1 Alfred W. Crosby, "The Past and Present of Environmental History," *American Historical Review* 100, no. 4 (1995): 1188.

2 Margaret Jacob, "Thinking Unfashionable Thoughts, Asking Unfashionable Questions," *American Historical Review* 105 (2000): 495.

WHERE NEXT?

KEY POINTS

- Alfred W. Crosby's work will remain a foundational text in environmental history.
- The text continues to influence historians who seek to examine the impact non-human actors have had on the course of history.
- The "Columbian exchange" continues to be used as a framework for conceiving of historical periods.

Potential

If Alfred W. Crosby's *The Columbian Exchange: Biological and Cultural Consequences of 1492* must be considered as a foundational text, its continuing relevance, in the sense of its currency, is probably not assured. The ideas Crosby raises and the arguments he advances have had, and will continue to have, an enduring impact on historiography.* Today, chairs are endowed in environmental history and many historians consider the relationship between the human and non-human world. Crosby is in part responsible for this state of affairs.

Yet Crosby's 1972 work was so successful that many of the conversations it provoked have moved far beyond its scope. Crosby's relatively abstract and imprecise arguments about the impact of New World* foods on European population growth have been engaged more directly, with a growing body of work seeking to quantify that impact.[1] Similar conversations are taking place about the relationship between social structure and the course of epidemic disease. Even if the scholarship has moved far beyond *The Columbian Exchange*, one can discern Crosby's influence in each debate and discussion.

Furthermore, it is likely that, even though *The Columbian Exchange* will eventually be considered to be a text of historical importance rather than a "living" text, it will continue to be relevant to study. Few works since have synthesized such a variety of literatures into a coherent historical argument. In this sense the book is important to subsequent historiography and it can be expected to enjoy an enduring legacy as a masterful example of historical work.

> *"Thanks to the acceptance of Crosby's work, the term 'Columbian exchange' is now widely used to describe the complex and many-faceted chain of ecological exchanges and impacts that began with Columbus."*
>
> —— Louis De Vorsey, "European Encounters: Discovery and Exploration"

Future Directions

Scholars will continue to develop ideas that Crosby developed and originated. *The Columbian Exchange* discussed the movement of animals and plants between hemispheres, and recent work by Virginia DeJohn Anderson,* a scholar of early-American history, explicitly argues that animals themselves can colonize and that they served as powerful allies in European expansion.[2] While

Anderson's work focuses on colonial New England, the importance of animals in the history of colonization as a whole, in different regions of the world, appears to be an important direction for future research. Similarly, the effect of new-crop circulation continues to be hotly debated, and historians and economists will continue to quantify the impact.

The environmental historian J. R. McNeill* has also pushed Crosby's insights in new directions. In the foreword to the 2003 edition of *The Columbian Exchange*, McNeill recalled how the book influenced his thinking as a historian: "My first encounter with the book came on a rainy afternoon in 1982 when I picked it off of a shoulder-high shelf in an office I temporarily occupied. I read it in one gulp neglecting the possibility of supper." McNeill's 2010 book *Mosquito Empires* has applied the approach Crosby forged in *The Columbian Exchange* to provide an ecological answer to a question that has long troubled historians of the Caribbean: why were other Atlantic powers unable to dislodge Spain, a declining power, from the Caribbean in the seventeenth and eighteenth centuries?

Summary

In its arguments and execution, *The Columbian Exchange* stands out as one of the most important historical works of the past 50 years. Crosby persuasively argues that non-human factors caused the most significant consequences of Christopher Columbus's* 1492 voyage from Europe to the Americas. Although this idea is now so widely accepted as to appear patently obvious, it proved revolutionary when the book was first published in 1972. Furthermore, *The Columbian Exchange* remains unparalleled as an example of how historical inquiry might be conducted by drawing from the sciences in an interdisciplinary fashion.

Crosby engaged with scientific literature of the time with consummate skill and creativity, revealing how much these fields can teach historians. In complementary fashion, he also encouraged historians to contribute their ideas to epidemiology,* biology, and the other sciences. As a consequence, Crosby's work deserves continued attention as a foundational text in environmental history and also because it remains one of the best examples of truly interdisciplinary scholarship.

The central ideas of *The Columbian Exchange* have proved to be influential both in their specific claims and in the greater whole. For example, Crosby turned the historical world upside down when he proposed that the decisive factor in European conquest of the New World was the spread of epidemic disease. This marked a significant departure from the work of earlier scholars, who put greater weight on European technology and the American peoples' "primitive" religious and political systems.

In demonstrating the effects of disease, Crosby undermined these older narratives, while simultaneously arguing for the importance of non-human factors such as disease, crops, and animals. Similarly, his emphasis on the framework of a "Columbian exchange" set the stage for the now-dominant paradigm for periodizing history (that is, of conceiving of historical periods). The impact of these ideas at the time of publication, and their subsequent influence in the decades since, secure the place of Crosby's text in the list of great historical works. Much like the crops that fed Europe's population explosion in the post-Columbus years, *The Columbian Exchange* has acted like a seed, growing to nourish a grateful population of historians, scholars, and thinkers.

---

1 Nathan Nunn and Nancy Qian, "The Columbian Exchange: A History of Disease, Food, and Ideas," *Journal of Economic Perspectives* 24, no. 2 (2010): 163–88.

2 Virginia DeJohn Anderson, *Creatures of Empire: How Domestic Animals Transformed Early America* (Oxford: Oxford University Press, 2006).

GLOSSARY OF TERMS

1. **American Revolution:** a war of independence waged by 13 American colonies against British rule from 1775 to 1783 that resulted in the formation of the United States.

2. **Aztecs:** a Native American people founded in about 1100 b.c.e. and first known as inhabitants of the valleys of central Mexico.

3. **Biodiversity:** the measure of the variety of species of organisms found on the planet.

4. **Biological homogeneity:** a process in which the variety of species in an environment is reduced.

5. **Black Power movement:** an American movement that reached prominence in the 1960s and sought to promote the collective interests of African Americans.

6. **Civil Rights movement:** a social movement in the United States of the 1950s and 1960s that aimed at abolishing racial discrimination and obtaining civil rights for African Americans.

7. **Colonization:** the act of settling in a territory away from one's place of origin and establishing political control there.

8. **Conquistador:** a term referring to New World explorers and soldiers of the Spanish Empire.The best-known conquistadors were Hernán Cortés (1485–1547), conqueror of the Aztecs, and Francisco Pizarro (1475–1541), conqueror of the Incan empire.

9. **Demography:** the statistical study of human populations using statistics of births, deaths, and diseases.

10. **Epidemiology:** the study of the dynamics of disease epidemics.

11. **Eurocentrism:** the belief in the cultural exceptionalism and superiority of Euro-American civilizations.

12. **Historiography:** the study of methods historians use; the evolution of history as a discipline.

13. **Holistic:** an approach that relates to the study of the whole, as opposed to

separate parts.

14. **Imperialism:** a policy of expanding a country's influence through colonization and military force.

15. **Incas:** a pre-Columbian civilization centered in what is now Peru.

16. **Influenza pandemic of 1918:** a severe influenza pandemic lasting from January 1918 to December 1920 that infected nearly 500 million people worldwide and killed between 50 and 100 million.

17. **New World:** a term originating in the early sixteenth century, referring to the landmasses in the Western hemisphere on which European explorers made landfall.

18. **Old World:** the continents of Africa, Europe, and Asia, collectively regarded as the known world prior to European exploration in the Americas in the fifteenth and sixteenth centuries.

19. **Pathogens:** an infectious biological agent that causes diseases and illness to a host.

20. **Possibilism:** a theory that the environment sets limits, but does not determine, the social and cultural development of human societies.

21. **Smallpox:** an infectious disease caused by two viruses, *variola major* and *variola minor*.

22. **Spanish conquest:** a process of expansion of the Spanish monarchy into South America that began in 1492 and lasted for three centuries.

23. **Syphilis:** a disease transmitted by sexual contact. Its various symptoms include sores on the skin and impaired brain function.

24. **Vietnam War:** a protracted conflict (1954–75) in which communist North Vietnam fought against the government of South Vietnam and its principal ally, the United States.

25. **War of 1812:** a conflict lasting from 1812 to 1815 between the United States and Great Britain, partly caused by British attempts to restrict US trade.

PEOPLE MENTIONED IN THE TEXT

1. **José de Acosta (1539–1600)** was a Spanish Jesuit missionary and naturalist, known for his pioneering work on the natural history of South America.

2. **Virginia De John Anderson (b. 1947)** is a professor of early-American history at the University of Colorado, Boulder. She is known for her 2006 book, *Creatures of Empire: How Domestic Animals Transformed Early America*.

3. **Rachel Carson (1907–64)** was an American marine biologist and conservationist best known for her 1962 book *Silent Spring*, which explored the environmental effects of pesticide use.

4. **Pierre Chaunu (1923–2009)** was a French historian who specialized in Latin American history. He also studied French social and religious history of the sixteenth to eighteenth centuries.

5. **Frederic E. Clements (1874–1945)** was an American plant ecologist and pioneer in the study of plant evolution. He is best known for his theory of "climax community," which argues that, after ecological disturbance, vegetation tends naturally towards a mature climax state.

6. **Christopher Columbus (1451–1506)** was an Italian explorer who completed four voyages across the Atlantic under the auspices of the Spanish monarchy, and initiated Spanish colonization of the New World.

7. **William Cronon (b. 1954)** is a professor of history, geography, and environmental studies at the University of Madison-Wisconsin. He is best known for *Nature's Metropolis: Chicago and the Great West*, winner of the 1992 Bancroft Prize.

8. **Jared Diamond (b. 1937)** is a geographer at UCLA, best known for his work *Guns, Germs and Steel* (1997).

9. **Gary S. Dunbar (1931–2015)** was an American geographer at UCLA, best known for his work on the disciplinary history of nineteenth- and twentieth-century geography.

10. **Charles Elton (1900–91)** was an English zoologist and animal ecologist.

11. **Huayna Capac (1464–1527)** was an Incan emperor who died of smallpox following the arrival of the Spanish in South America in the early sixteenth

century.

12. **Margaret Jacob (b. 1943)** is a distinguished professor of history at UCLA, whose recent work has focused on the cultural origins of the Industrial Revolution.

13. **David Landes (1924–2013)** was an American economic historian best known for his work on the Industrial Revolution.

14. **Aldo Leopold (1887–1948)** was an American conservationist and advocate for wilderness protection.

15. **Patricia Limerick (b. 1951)** is an American environmental historian and one of the leading historians of the American West.

16. **J. R. McNeill (b. 1954)** is an environmental historian and historian at Georgetown University, best known for his 2000 book *Something New Under the Sun: An Environmental History of the Twentieth-Century World.*

17. **Robert E. Moody (1901–83)** was an American historian who spent the majority of his career at Boston University, specializing in early American history.

18. **Nathan Nunn** is a native of Canada and a professor of economics at Harvard University. His work focuses on economic history and developmental economics.

19. **Nancy Qian** is an associate professor of economics at Yale University who has worked on the history of famine and economic development.

20. **Frederick Jackson Turner (1861–1932)** was an American historian of the early twentieth century best known for his writings on the American frontier.

21. **Charles Verlinden (1907–96)** was a Belgian medieval historian who specialized in economic history and the history of colonialism.

22. **Paul Vidal de la Blache (1845–1918)** was one of the founders of French geography, best known for his theory of "possibilism," which opposed environmental determinism (that is, it proposed that the environment was not the ultimate decider in the shape of human societies).

23. **Walter Prescott Webb (1888–1963)** was an American historian best known for his groundbreaking work on the environmental history of the American West.

24. **Richard White (b. 1947)** is an American environmental historian at Stanford University, known for his 1983 work *The Roots of Dependency: Subsistence, Environment and Social Change Among the Choctaws, Pawnees and Navajos.*

25. **Donald Worster (b. 1941)** is a distinguished professor of American history at the University of Kansas. One of the founders of environmental history, he is best known for his 1979 book *Dust Bowl: The Southern Plains in the 1930s.*

 **WORKS CITED**

1. Anderson, Virginia De John. *Creatures of Empire: How Domestic Animals Transformed Early America*. Oxford: Oxford University Press, 2006.

2. Carson, Rachel. *Silent Spring*. New York: Houghton Mifflin, 1962.

3. Cronon, William. *Changes in the Land: Indians, Colonists and the Ecology of New England*. New York: Hill and Wang, 1983 (revised edn 2003).

4. Crosby, Alfred W. *America, Russia, Hemp, and Napoleon: American trade with Russia and the Baltic, 1783–1812*. Columbus: Ohio State University Press, 1965.

5. "Virgin Soil Epidemics as a Factor in the Aboriginal Depopulation in America." *William and Mary Quarterly* 33, no. 2 (1976): 289–99.

6. *The Columbian Voyages, the Columbian Exchange, and Their Historians*. Washington, DC: American Historical Association, 1987.

7. "Reassessing 1492." *American Quarterly* 41, no. 4 (1989): 661–9.

8. "The Past and Present of Environmental History." *American Historical Review* 100, no. 4 (1995): 1177–89.

9. *America's Forgotten Pandemic: The Influenza of 1918*. 2nd edn. Cambridge: Cambridge University Press, 2003.

10. *The Columbian Exchange: Biological and Cultural Consequences of 1492*. Westport, CT: Praeger, 2003.

11. *Ecological Imperialism: The Biological Expansion of Europe, 900–1900*. 2nd edn. Cambridge: Cambridge University Press, 2004.

12. Diamond, Jared M. *Guns, Germs and Steel: A Short History of Everybody for the Last 13,000 Years*. New York: Vintage, 1998.

13. Dunbar, G. S. "Review of *The Columbian Exchange*." *William and Mary Quarterly* 30, no. 3 (1973): 542–3.

14. Jacob, Margaret. "Thinking Unfashionable Thoughts, Asking Unfashionable Questions." *American Historical Review* 105 (2000): 495.

15. Landes, David S. *The Wealth and Poverty of Nations: Why Some Are So Rich and Some So Poor*. New York: W.W. Norton, 1998.

16. McNeill, J. R. Foreword to *The Columbian Exchange: Biological and Cultural Consequences of 1492*, by Alfred W. Crosby. Westport, CT: Praeger, 2003.

17. *Mosquito Empires: Ecology and War in the Greater Caribbean, 1620–1914*. Cambridge: Cambridge University Press, 2010.

18. Nunn, Nathan, and Nancy Qian. "The Columbian Exchange: A History of Disease, Food, and Ideas." *Journal of Economic Perspectives* 24, no. 2 (2010): 163–88.

19. Turner, Frederick Jackson. "The Significance of the Frontier in American History." In *The Frontier in American History.* New York: Henry Holt, 1921.

20. Webb, Walter Prescott. *The Great Plains.* Lincoln: University of Nebraska Press, 1981.

21. White, Richard. *The Roots of Dependency: Subsistence, Environment and Social Change Among the Choctaws, Pawnees and Navajos.* Lincoln: University of Nebraska Press, 1983.

22. *The Organic Machine: The Remaking of the Columbia River.* New York: Hill and Wang, 1995.

原书作者简介

艾尔弗雷德·W.克罗斯比，1931 年出生于美国波士顿，哈佛大学教育学院教学艺术硕士，波士顿大学历史学博士。然而，克罗斯比很快对基于"世界将越变越好""好人总是获胜"等观点的传统历史研究方法提出质疑。随着他对环境问题，以及民权运动、越南战争等政治事件的兴趣日益浓厚，这些疑虑也日渐加深。因此克罗斯比重新审视历史，并开辟出新的路径来解释这些事件发生的原因。

本书作者简介

约书亚·施佩希特博士，2014 年获哈佛大学历史学博士学位，研究方向为 19 世纪美洲牲畜贸易的环境史。他目前任莫纳什大学历史专业讲师。

艾蒂安·斯托克兰德，哥伦比亚大学环境史专业在读博士生。

世界名著中的批判性思维

《世界思想宝库钥匙丛书》致力于深入浅出地阐释全世界著名思想家的观点，不论是谁、在何处都能了解到，从而推进批判性思维发展。

《世界思想宝库钥匙丛书》与世界顶尖大学的一流学者合作，为一系列学科中最有影响的著作推出新的分析文本，介绍其观点和影响。在这一不断扩展的系列中，每种选入的著作都代表了历经时间考验的思想典范。通过为这些著作提供必要背景、揭示原作者的学术渊源以及说明这些著作所产生的影响，本系列图书希望让读者以新视角看待这些划时代的经典之作。读者应学会思考、参与并挑战这些著作中的观点，而不是简单接受它们。

ABOUT THE AUTHOR OF THE ORIGINAL WORK

Alfred W. Crosby was born in Boston in the United States in 1931 and received his Masters in the art of teaching from the Harvard School of Education and a PhD in history from Boston University. However, he soon became skeptical of the ways in history was taught based on the view that "the world became progressively better" and that "the good guys always won." These doubts were fostered by a growing personal interest in both environmental matters and political issues of the day, such as the civil rights movement and the Vietnam War. Crosby reinvestigated history and developed new ways of explaining why events happened as they did.

ABOUT THE AUTHORS OF THE ANALYSIS

Dr Joshua Specht completed his PhD in history at Harvard in 2014, working on the environmental history of the cattle trade in nineteenth-century America. He is currently a lecturer in history at Monash University.
Etienne Stockland is researching a PhD in environmental history at Columbia University.

ABOUT MACAT
GREAT WORKS FOR CRITICAL THINKING

Macat is focused on making the ideas of the world's great thinkers accessible and comprehensible to everybody, everywhere, in ways that promote the development of enhanced critical thinking skills.

It works with leading academics from the world's top universities to produce new analyses that focus on the ideas and the impact of the most influential works ever written across a wide variety of academic disciplines. Each of the works that sit at the heart of its growing library is an enduring example of great thinking. But by setting them in context — and looking at the influences that shaped their authors, as well as the responses they provoked — Macat encourages readers to look at these classics and game-changers with fresh eyes. Readers learn to think, engage and challenge their ideas, rather than simply accepting them.

批判性思维与《哥伦布大交换》

首要批判性思维技巧：创造性思维

次要批判性思维技巧：评估性思维

　　一种对历史学的批评是，历史学家常孤立地进行研究而未能利用其他学科学者的模型和证据。但这样的批评不应针对艾尔弗雷德·克罗斯比。他的著作《哥伦布大交换》吸收了对硬科学、人类学和地理学广泛阅读的成果，居于环境历史学研究的奠基著作之列。

　　从这个意义上来讲，克罗斯比的这部典型著作无疑是创造性思维的好例子。该书建立起新的关联，认为欧洲对新世界殖民统治的成功，更多是生物灾难的产物（以引入新疾病的方式），与人类行为的关系不大。该书还认为，新、旧世界接触的最重要影响并不关乎政治，如新帝国的建立，而是关乎文化和饮食，例如，由于从新世界引入粮食作物，中国人口增加了两倍。

　　克罗斯比在《哥伦布大交换》中提出的中心假设极富刺激性和影响力，罕有原创性假设能够与之比肩。

CRITICAL THINKING AND *THE COLUMBIAN EXCHANGE*

- Primary critical thinking skill: CREATIVE THINKING
- Secondary critical thinking skill: EVALUATION

One criticism of history is that historians all too often study in isolation, failing to take advantage of models and evidence from scholars in other disciplines. This is not a charge that can be laid at the door of Alfred Crosby. His book *The Columbian Exchange* not only incorporates the results of wide reading in the hard sciences, anthropology and geography, but also stands as one of the foundation stones of the study of environmental history.

In this sense, Crosby's defining work is undoubtedly a fine example of creative thinking. The book comes up with new connections that explain the European success in colonizing the New World more as the product of biological catastrophe (in the shape of the introduction of new diseases) than of the actions of men, and posits that the most important consequences of contact were not political—the establishment of new empires—but cultural and culinary; the population of China tripled, for example, as the result of the introduction of crops from the New World.

Few original hypotheses have proved as stimulating or as influential as the one that Crosby places at the heart of *The Columbian Exchange.*

《世界思想宝库钥匙丛书》简介

《世界思想宝库钥匙丛书》致力于为一系列在各领域产生重大影响的人文社科类经典著作提供独特的学术探讨。每一本读物都不仅仅是原经典著作的内容摘要，而是介绍并深入研究原经典著作的学术渊源、主要观点和历史影响。这一丛书的目的是提供一套学习资料，以促进读者掌握批判性思维，从而更全面、深刻地去理解重要思想。

每一本读物分为 3 个部分：学术渊源、学术思想和学术影响，每个部分下有 4 个小节。这些章节旨在从各个方面研究原经典著作及其反响。

由于独特的体例，每一本读物不但易于阅读，而且另有一项优点：所有读物的编排体例相同，读者在进行某个知识层面的调查或研究时可交叉参阅多本该丛书中的相关读物，从而开启跨领域研究的路径。

为了方便阅读，每本读物最后还列出了术语表和人名表（在书中则以星号 * 标记），此外还有参考文献。

《世界思想宝库钥匙丛书》与剑桥大学合作，理清了批判性思维的要点，即如何通过 6 种技能来进行有效思考。其中 3 种技能让我们能够理解问题，另 3 种技能让我们有能力解决问题。这 6 种技能合称为"批判性思维 PACIER 模式"，它们是：

分析：了解如何建立一个观点；

评估：研究一个观点的优点和缺点；

阐释：对意义所产生的问题加以理解；

创造性思维：提出新的见解，发现新的联系；

解决问题：提出切实有效的解决办法；

理性化思维：创建有说服力的观点。

了解更多信息，请浏览 www.macat.com。

THE MACAT LIBRARY

The Macat Library is a series of unique academic explorations of seminal works in the humanities and social sciences — books and papers that have had a significant and widely recognised impact on their disciplines. It has been created to serve as much more than just a summary of what lies between the covers of a great book. It illuminates and explores the influences on, ideas of, and impact of that book. Our goal is to offer a learning resource that encourages critical thinking and fosters a better, deeper understanding of important ideas.

Each publication is divided into three Sections: Influences, Ideas, and Impact. Each Section has four Modules. These explore every important facet of the work, and the responses to it.

This Section-Module structure makes a Macat Library book easy to use, but it has another important feature. Because each Macat book is written to the same format, it is possible (and encouraged!) to cross-reference multiple Macat books along the same lines of inquiry or research. This allows the reader to open up interesting interdisciplinary pathways.

To further aid your reading, lists of glossary terms and people mentioned are included at the end of this book (these are indicated by an asterisk [*] throughout) — as well as a list of works cited.

Macat has worked with the University of Cambridge to identify the elements of critical thinking and understand the ways in which six different skills combine to enable effective thinking.

Three allow us to fully understand a problem; three more give us the tools to solve it. Together, these six skills make up the PACIER model of critical thinking. They are:

ANALYSIS — understanding how an argument is built
EVALUATION — exploring the strengths and weaknesses of an argument
INTERPRETATION — understanding issues of meaning
CREATIVE THINKING — coming up with new ideas and fresh connections
PROBLEM-SOLVING — producing strong solutions
REASONING — creating strong arguments

To find out more, visit WWW.MACAT.COM.

"《世界思想宝库钥匙丛书》提供了独一无二的跨学科学习和研究工具。它介绍那些革新了各自学科研究的经典著作，还邀请全世界一流专家和教育机构进行严谨的分析，为每位读者打开世界顶级教育的大门。"

—— 安德烈亚斯·施莱歇尔，
经济合作与发展组织教育与技能司司长

"《世界思想宝库钥匙丛书》直面大学教育的巨大挑战……他们组建了一支精干而活跃的学者队伍，来推出在研究广度上颇具新意的教学材料。"

—— 布罗尔斯教授、勋爵，剑桥大学前校长

"《世界思想宝库钥匙丛书》的愿景令人赞叹。它通过分析和阐释那些曾深刻影响人类思想以及社会、经济发展的经典文本，提供了新的学习方法。它推动批判性思维，这对于任何社会和经济体来说都是至关重要的。这就是未来的学习方法。"

—— 查尔斯·克拉克阁下，英国前教育大臣

"对于那些影响了各自领域的著作，《世界思想宝库钥匙丛书》能让人们立即了解到围绕那些著作展开的评论性言论，这让该系列图书成为在这些领域从事研究的师生们不可或缺的资源。"

—— 威廉·特朗佐教授，加利福尼亚大学圣地亚哥分校

"Macat offers an amazing first-of-its-kind tool for interdisciplinary learning and research. Its focus on works that transformed their disciplines and its rigorous approach, drawing on the world's leading experts and educational institutions, opens up a world-class education to anyone."

—— Andreas Schleicher, Director for Education and Skills, Organisation for Economic Co-operation and Development

"Macat is taking on some of the major challenges in university education... They have drawn together a strong team of active academics who are producing teaching materials that are novel in the breadth of their approach."

—— Prof Lord Broers, former Vice-Chancellor of the University of Cambridge

"The Macat vision is exceptionally exciting. It focuses upon new modes of learning which analyse and explain seminal texts which have profoundly influenced world thinking and so social and economic development. It promotes the kind of critical thinking which is essential for any society and economy. This is the learning of the future."

—— Rt Hon Charles Clarke, former UK Secretary of State for Education

"The Macat analyses provide immediate access to the critical conversation surrounding the books that have shaped their respective discipline, which will make them an invaluable resource to all of those, students and teachers, working in the field."

—— Prof William Tronzo, University of California at San Diego

TITLE	中文书名	类别
An Analysis of Arjun Appadurai's *Modernity at Large: Cultural Dimensions of Globalisation*	解析阿尔君·阿帕杜莱《消失的现代性：全球化的文化维度》	人类学
An Analysis of Claude Lévi-Strauss's *Structural Anthropology*	解析克劳德·列维-斯特劳斯《结构人类学》	人类学
An Analysis of Marcel Mauss's *The Gift*	解析马塞尔·莫斯《礼物》	人类学
An Analysis of Jared M. Diamond's *Guns, Germs, and Steel: The Fate of Human Societies*	解析贾雷德·戴蒙德《枪炮、病菌与钢铁：人类社会的命运》	人类学
An Analysis of Clifford Geertz's *The Interpretation of Cultures*	解析克利福德·格尔茨《文化的解释》	人类学
An Analysis of Philippe Ariès's *Centuries of Childhood: A Social History of Family Life*	解析菲利浦·阿利埃斯《儿童的世纪：旧制度下的儿童和家庭生活》	人类学
An Analysis of W. Chan Kim & Renée Mauborgne's *Blue Ocean Strategy*	解析金伟灿/勒妮·莫博涅《蓝海战略》	商业
An Analysis of John P. Kotter's *Leading Change*	解析约翰·P.科特《领导变革》	商业
An Analysis of Michael E. Porter's *Competitive Strategy: Creating and Sustaining Superior Performance*	解析迈克尔·E.波特《竞争战略：分析产业和竞争对手的技术》	商业
An Analysis of Jean Lave & Etienne Wenger's *Situated Learning: Legitimate Peripheral Participation*	解析琼·莱夫/艾蒂纳·温格《情境学习：合法的边缘性参与》	商业
An Analysis of Douglas McGregor's *The Human Side of Enterprise*	解析道格拉斯·麦格雷戈《企业的人性面》	商业
An Analysis of Milton Friedman's *Capitalism and Freedom*	解析米尔顿·弗里德曼《资本主义与自由》	商业
An Analysis of Ludwig von Mises's *The Theory of Money and Credit*	解析路德维希·冯·米塞斯《货币和信用理论》	经济学
An Analysis of Adam Smith's *The Wealth of Nations*	解析亚当·斯密《国富论》	经济学
An Analysis of Thomas Piketty's *Capital in the Twenty-First Century*	解析托马斯·皮凯蒂《21世纪资本论》	经济学
An Analysis of Nassim Nicholas Taleb's *The Black Swan: The Impact of the Highly Improbable*	解析纳西姆·尼古拉斯·塔勒布《黑天鹅：如何应对不可预知的未来》	经济学
An Analysis of Ha-Joon Chang's *Kicking Away the Ladder*	解析张夏准《富国陷阱：发达国家为何踢开梯子》	经济学
An Analysis of Thomas Robert Malthus's *An Essay on the Principle of Population*	解析托马斯·马尔萨斯《人口论》	经济学

An Analysis of John Maynard Keynes's *The General Theory of Employment, Interest and Money*	解析约翰·梅纳德·凯恩斯《就业、利息和货币通论》	经济学
An Analysis of Milton Friedman's *The Role of Monetary Policy*	解析米尔顿·弗里德曼《货币政策的作用》	经济学
An Analysis of Burton G. Malkiel's *A Random Walk Down Wall Street*	解析伯顿·G. 马尔基尔《漫步华尔街》	经济学
An Analysis of Friedrich A. Hayek's *The Road to Serfdom*	解析弗里德里希·A. 哈耶克《通往奴役之路》	经济学
An Analysis of Charles P. Kindleberger's *Manias, Panics, and Crashes: A History of Financial Crises*	解析查尔斯·P. 金德尔伯格《疯狂、惊恐和崩溃：金融危机史》	经济学
An Analysis of Amartya Sen's *Development as Freedom*	解析阿马蒂亚·森《以自由看待发展》	经济学
An Analysis of Rachel Carson's *Silent Spring*	解析蕾切尔·卡森《寂静的春天》	地理学
An Analysis of Charles Darwin's *On the Origin of Species: by Means of Natural Selection, or The Preservation of Favoured Races in the Struggle for Life*	解析查尔斯·达尔文《物种起源》	地理学
An Analysis of World Commission on Environment and Development's *The Brundtland Report, Our Common Future*	解析世界环境与发展委员会《布伦特兰报告：我们共同的未来》	地理学
An Analysis of James E. Lovelock's *Gaia: A New Look at Life on Earth*	解析詹姆斯·E. 拉伍洛克《盖娅：地球生命的新视野》	地理学
An Analysis of Paul Kennedy's *The Rise and Fall of the Great Powers: Economic Change and Military Conflict from 1500—2000*	解析保罗·肯尼迪《大国的兴衰：1500—2000 年的经济变革与军事冲突》	历史
An Analysis of Janet L. Abu-Lughod's *Before European Hegemony: The World System A. D. 1250—1350*	解析珍妮特·L. 阿布-卢格霍德《欧洲霸权之前：1250—1350 年的世界体系》	历史
An Analysis of Alfred W. Crosby's *The Columbian Exchange: Biological and Cultural Consequences of 1492*	解析艾尔弗雷德·W. 克罗斯比《哥伦布大交换：1492 年以后的生物影响和文化冲击》	历史
An Analysis of Tony Judt's *Postwar: A History of Europe since 1945*	解析托尼·贾德《战后欧洲史》	历史
An Analysis of Richard J. Evans's *In Defence of History*	解析理查德·J. 艾文斯《捍卫历史》	历史
An Analysis of Eric Hobsbawm's *The Age of Revolution: Europe 1789–1848*	解析艾瑞克·霍布斯鲍姆《革命的年代：欧洲 1789—1848 年》	历史

An Analysis of Roland Barthes's *Mythologies*	解析罗兰·巴特《神话学》	文学与批判理论
An Analysis of Simon de Beauvoir's *The Second Sex*	解析西蒙娜·德·波伏娃《第二性》	文学与批判理论
An Analysis of Edward W. Said's *Orientalism*	解析爱德华·W. 萨义德《东方主义》	文学与批判理论
An Analysis of Virginia Woolf's *A Room of One's Own*	解析弗吉尼亚·伍尔芙《一间自己的房间》	文学与批判理论
An Analysis of Judith Butler's *Gender Trouble*	解析朱迪斯·巴特勒《性别麻烦》	文学与批判理论
An Analysis of Ferdinand de Saussure's *Course in General Linguistics*	解析费尔迪南·德·索绪尔《普通语言学教程》	文学与批判理论
An Analysis of Susan Sontag's *On Photography*	解析苏珊·桑塔格《论摄影》	文学与批判理论
An Analysis of Walter Benjamin's *The Work of Art in the Age of Mechanical Reproduction*	解析瓦尔特·本雅明《机械复制时代的艺术作品》	文学与批判理论
An Analysis of W.E.B. Du Bois's *The Souls of Black Folk*	解析 W.E.B. 杜博伊斯《黑人的灵魂》	文学与批判理论
An Analysis of Plato's *The Republic*	解析柏拉图《理想国》	哲学
An Analysis of Plato's *Symposium*	解析柏拉图《会饮篇》	哲学
An Analysis of Aristotle's *Metaphysics*	解析亚里士多德《形而上学》	哲学
An Analysis of Aristotle's *Nicomachean Ethics*	解析亚里士多德《尼各马可伦理学》	哲学
An Analysis of Immanuel Kant's *Critique of Pure Reason*	解析伊曼努尔·康德《纯粹理性批判》	哲学
An Analysis of Ludwig Wittgenstein's *Philosophical Investigations*	解析路德维希·维特根斯坦《哲学研究》	哲学
An Analysis of G.W.F. Hegel's *Phenomenology of Spirit*	解析 G.W.F. 黑格尔《精神现象学》	哲学
An Analysis of Baruch Spinoza's *Ethics*	解析巴鲁赫·斯宾诺莎《伦理学》	哲学
An Analysis of Hannah Arendt's *The Human Condition*	解析汉娜·阿伦特《人的境况》	哲学
An Analysis of G.E.M. Anscombe's *Modern Moral Philosophy*	解析 G.E.M. 安斯康姆《现代道德哲学》	哲学
An Analysis of David Hume's *An Enquiry Concerning Human Understanding*	解析大卫·休谟《人类理解研究》	哲学

An Analysis of Søren Kierkegaard's *Fear and Trembling*	解析索伦·克尔凯郭尔《恐惧与战栗》	哲学
An Analysis of René Descartes's *Meditations on First Philosophy*	解析勒内·笛卡尔《第一哲学沉思录》	哲学
An Analysis of Friedrich Nietzsche's *On the Genealogy of Morality*	解析弗里德里希·尼采《论道德的谱系》	哲学
An Analysis of Gilbert Ryle's *The Concept of Mind*	解析吉尔伯特·赖尔《心的概念》	哲学
An Analysis of Thomas Kuhn's *The Structure of Scientific Revolutions*	解析托马斯·库恩《科学革命的结构》	哲学
An Analysis of John Stuart Mill's *Utilitarianism*	解析约翰·斯图亚特·穆勒《功利主义》	哲学
An Analysis of Aristotle's *Politics*	解析亚里士多德《政治学》	政治学
An Analysis of Niccolò Machiavelli's *The Prince*	解析尼科洛·马基雅维利《君主论》	政治学
An Analysis of Karl Marx's *Capital*	解析卡尔·马克思《资本论》	政治学
An Analysis of Benedict Anderson's *Imagined Communities*	解析本尼迪克特·安德森《想象的共同体》	政治学
An Analysis of Samuel P. Huntington's *The Clash of Civilizations and the Remaking of World Order*	解析塞缪尔·P.亨廷顿《文明的冲突与世界秩序重建》	政治学
An Analysis of Alexis de Tocqueville's *Democracy in America*	解析阿列克西·德·托克维尔《论美国的民主》	政治学
An Analysis of J. A. Hobson's *Imperialism: A Study*	解析约·阿·霍布森《帝国主义》	政治学
An Analysis of Thomas Paine's *Common Sense*	解析托马斯·潘恩《常识》	政治学
An Analysis of John Rawls's *A Theory of Justice*	解析约翰·罗尔斯《正义论》	政治学
An Analysis of Francis Fukuyama's *The End of History and the Last Man*	解析弗朗西斯·福山《历史的终结与最后的人》	政治学
An Analysis of John Locke's *Two Treatises of Government*	解析约翰·洛克《政府论》	政治学
An Analysis of Sun Tzu's *The Art of War*	解析孙武《孙子兵法》	政治学
An Analysis of Henry Kissinger's *World Order: Reflections on the Character of Nations and the Course of History*	解析亨利·基辛格《世界秩序》	政治学
An Analysis of Jean-Jacques Rousseau's *The Social Contract*	解析让-雅克·卢梭《社会契约论》	政治学

An Analysis of Odd Arne Westad's *The Global Cold War: Third World Interventions and the Making of Our Times*	解析文安立《全球冷战：美苏对第三世界的干涉与当代世界的形成》	政治学
An Analysis of Sigmund Freud's *The Interpretation of Dreams*	解析西格蒙德·弗洛伊德《梦的解析》	心理学
An Analysis of William James' *The Principles of Psychology*	解析威廉·詹姆斯《心理学原理》	心理学
An Analysis of Philip Zimbardo's *The Lucifer Effect*	解析菲利普·津巴多《路西法效应》	心理学
An Analysis of Leon Festinger's *A Theory of Cognitive Dissonance*	解析利昂·费斯汀格《认知失调论》	心理学
An Analysis of Richard H. Thaler & Cass R. Sunstein's *Nudge: Improving Decisions about Health, Wealth, and Happiness*	解析理查德·H.泰勒／卡斯·R.桑斯坦《助推：如何做出有关健康、财富和幸福的更优决策》	心理学
An Analysis of Gordon Allport's *The Nature of Prejudice*	解析高尔登·奥尔波特《偏见的本质》	心理学
An Analysis of Steven Pinker's *The Better Angels of Our Nature: Why Violence Has Declined*	解析斯蒂芬·平克《人性中的善良天使：暴力为什么会减少》	心理学
An Analysis of Stanley Milgram's *Obedience to Authority*	解析斯坦利·米尔格拉姆《对权威的服从》	心理学
An Analysis of Betty Friedan's *The Feminine Mystique*	解析贝蒂·弗里丹《女性的奥秘》	心理学
An Analysis of David Riesman's *The Lonely Crowd: A Study of the Changing American Character*	解析大卫·理斯曼《孤独的人群：美国人社会性格演变之研究》	社会学
An Analysis of Franz Boas's *Race, Language and Culture*	解析弗朗兹·博厄斯《种族、语言与文化》	社会学
An Analysis of Pierre Bourdieu's *Outline of a Theory of Practice*	解析皮埃尔·布尔迪厄《实践理论大纲》	社会学
An Analysis of Max Weber's *The Protestant Ethic and the Spirit of Capitalism*	解析马克斯·韦伯《新教伦理与资本主义精神》	社会学
An Analysis of Jane Jacobs's *The Death and Life of Great American Cities*	解析简·雅各布斯《美国大城市的死与生》	社会学
An Analysis of C. Wright Mills's *The Sociological Imagination*	解析C.赖特·米尔斯《社会学的想象力》	社会学
An Analysis of Robert E. Lucas Jr.'s *Why Doesn't Capital Flow from Rich to Poor Countries?*	解析小罗伯特·E.卢卡斯《为何资本不从富国流向穷国？》	社会学

An Analysis of Émile Durkheim's *On Suicide*	解析埃米尔·迪尔凯姆《自杀论》	社会学
An Analysis of Eric Hoffer's *The True Believer: Thoughts on the Nature of Mass Movements*	解析埃里克·霍弗《狂热分子：群众运动圣经》	社会学
An Analysis of Jared M. Diamond's *Collapse: How Societies Choose to Fail or Survive*	解析贾雷德·M.戴蒙德《大崩溃：社会如何选择兴亡》	社会学
An Analysis of Michel Foucault's *The History of Sexuality Vol. 1: The Will to Knowledge*	解析米歇尔·福柯《性史（第一卷）：求知意志》	社会学
An Analysis of Michel Foucault's *Discipline and Punish*	解析米歇尔·福柯《规训与惩罚》	社会学
An Analysis of Richard Dawkins's *The Selfish Gene*	解析理查德·道金斯《自私的基因》	社会学
An Analysis of Antonio Gramsci's *Prison Notebooks*	解析安东尼奥·葛兰西《狱中札记》	社会学
An Analysis of Augustine's *Confessions*	解析奥古斯丁《忏悔录》	神学
An Analysis of C. S. Lewis's *The Abolition of Man*	解析C.S.路易斯《人之废》	神学

图书在版编目（CIP）数据

解析艾尔弗雷德·W. 克罗斯比《哥伦布大交换：1492年以后的生物影响和
文化冲击》：汉、英 / 约书亚·施佩希特（Joshua Specht），艾蒂安·斯托克兰
德（Etienne Stockland）著；宫昀译. —上海：上海外语教育出版社，2019
（世界思想宝库钥匙丛书）
ISBN 978-7-5446-5839-3

Ⅰ. ①解… Ⅱ. ①约… ②艾… ③宫… Ⅲ. ①生物地理学—研究—汉、英②社
会发展—影响—自然环境—研究—汉、英 Ⅳ. ①Q15②X24

中国版本图书馆CIP数据核字（2019）第076629号

This Chinese-English bilingual edition of *An Analysis of Alfred W. Crosby's* The Columbian
Exchange is published by arrangement with MACAT International Limited.
Licensed for sale throughout the world.

本书汉英双语版由Macat国际有限公司授权上海外语教育出版社有限公司出版。
供在全世界范围内发行、销售。

图字：09 - 2018 - 549

出版发行：上海外语教育出版社
　　　　　　（上海外国语大学内）　邮编：200083
电　　话：021-65425300（总机）
电子邮箱：bookinfo@sflep.com.cn
网　　址：http://www.sflep.com
责任编辑：王叶涵

印　　刷：上海信老印刷厂
开　　本：890×1240　1/32　印张5.25　字数107千字
版　　次：2019年8月第1版　2019年8月第1次印刷
印　　数：2 100 册

书　　号：ISBN 978-7-5446-5839-3 / Q
定　　价：30.00 元
　　　本版图书如有印装质量问题，可向本社调换
　　　质量服务热线：4008-213-263　电子邮箱：**editorial@sflep.com**